Topics in Atmospheric and Oceanographic Sciences

Editors: Michael Ghil   Robert Sadourny   Jürgen Sündermann

Problems and Prospects in Long and Medium Range Weather Forecasting

Edited by
D. M. Burridge and E. Källén

With 74 Figures

Springer-Verlag
Berlin Heidelberg New York Tokyo 1984

Dr. DAVID M. BURRIDGE, European Centre for Medium Range,
Weather Forecasting, Shinfield Park, Reading, Berkshire RG2 9AX, England

Dr. ERLAND KÄLLÉN, Department of Meteorology,
University of Stockholm, Arrhenius Laboratoriet, S-10691 Stockholm, Sweden

Series Editors:

Dr. MICHAEL GHIL, Courant Institute of Mathematical Sciences,
New York University, 251 Mercer Street, New York, NY 10012 / USA

Dr. ROBERT SADOURNY, Laboratoire de Météorologie Dynamique,
Ecole Normale Supérieure, 24 rue Lhomond, 75231 Paris Cedex 05 / France

Dr. JÜRGEN SÜNDERMANN, Universität Hamburg, Institut für Meereskunde,
Heimhuder Straße 71, 2000 Hamburg 13 / FRG

ISBN 3-540-12827-1 Springer-Verlag Berlin Heidelberg New York Tokyo
ISBN 0-387-12827-1 Springer-Verlag New York Heidelberg Berlin Tokyo

Library of Congress Cataloging in Publication Data. Main entry under title: Problems and prospects in long and medium range weather forecasting. (Topics in atmospheric and oceanographic sciences) 1. Weather forecasting. 2. Long-range weather forecasting. I. Burridge, D. M. II. Källén, Erland. III. Series. QC995.P884 1983 551.6'365 83-16813 ISBN 0-387-12827-1 (U.S.)

This work is subject copyright. All rights are reserved, whether the whole or part of the material is concerned, specifically those of translation, reprinting, re-use of illustrations, broadcasting, reproduction by photocopying machine or similar means, and storage in data banks. Under § 54 of the German Copyright Law, where copies are made for other than private use, a fee is payable to "Verwertungsgesellschaft Wort", Munich.

© by Springer-Verlag, Berlin Heidelberg 1984
Printed in Germany.

The use of registered names, trademarks, etc. in this publication does not imply, even in the absence of a specific statement, that such names are exempt from the relevant protective laws and regulations and therefore free for general use.

Printing and bookbinding: Beltz Offsetdruck, Hemsbach/Bergstr.
2132/3130-543210

Preface

Forecasting the weather for the long and medium range is a difficult and scientifically challenging problem. Since the first operational weather prediction by numerical methods was carried out (on the BESK computer in Stockholm, Sweden, 1954) there has been an ever accelerating development in computer technology. Hand in hand has followed a tremendous increase in the complexity of the atmospheric models used for weather prediction. The ability of these models to predict future states of the atmosphere has also increased rapidly, both due to model development and due to more accurate and plentiful observations of the atmosphere to define the initial state for model integrations. It may however be argued on theoretical grounds that even if we have an almost perfect model with almost perfect initial data, we will never be able to make an accurate weather prediction more than a few weeks ahead. This is due to the inherent instability of the atmosphere and work in this field was pioneered by E.Lorenz. It is generally referred to as atmospheric predictability and in the opening chapter of this book Professor Lorenz gives us an overview of the problem of atmospheric predictability.

The contributions to this book were originally presented at the 1981 ECMWF Seminar (ECMWF - European Centre for Medium Range Weather Forecasts) which was held at ECMWF in Reading, England, in September 1981. When the Centre was set up in 1975 its task was to develop a numerical model for predicting the weather for the medium range (3-10 days ahead) using the latest available computer technology and scientific methods. The first ECMWF Seminar in 1975 was devoted to the scientific foundation of medium range weather forecasting and the 1981 Seminar may be seen as a follow up to this first Seminar. The contribution by L. Bengtsson, the present Director, gives an historical review of the developments which led up to the design of a very successful forecasting system. At present operational medium-range forecasts are produced every day and are distributed world wide. There are, of course, still many problems remaining with the forecasting system and these aspects are covered in two contributions by A. Simmons and G. Cats. The first, by A. Simmons, deals with the model aspects while the second, by G. Cats, takes up some problems with, and examples of, model sensitivity to the initial state.

As we have already stated, there is a theoretical limit to the predictability of atmospheric flow and beyond this limit one can only hope to predict the largest scales and long time averages. Attempts to do this by using complex, deterministic general circulation models of the atmosphere are presented in two papers by J. Shukla. In the first he discusses the possibilities of using this approach and in particular in which geographical regions it can be expected that the method would be most successful. In his second contribution, Dr. Shukla discusses the sensitivity of a model to changes in the spatial boundary conditions. When verifying the results from model simulations of this type a thorough test of the statistical significance must be made. A chapter on statistical methods by C. Leith gives a theoretical treatment of some aspects of this problem.

Despite the fact that the predictability of the atmosphere is limited to a week or two on the average, we know that there are certain large-scale flow configurations which are persistent and thus potentially more predictable. One such flow configuration is the atmospheric blocking pattern, regularly observed over the Atlantic and Pacific Oceans. Two possible nonlinear mechanisms which may explain the stability and persistence of this flow pattern are covered in two contributions by E. Källén and C. Leith.

This book does not attempt to cover all aspects of long and medium range weather forecasting. We have only included some contributions from the 1981 ECMWF Seminars which we feel can be of general interest. In doing so we have concentrated on the numerical weather prediction method which has proven itself to be successful and practically useful. We have also tried to include a general background to the predictability problem and some theoretical ideas which may be useful for future long-range forecasting methods.

We are pleased to acknowledge Iris Rhodes for typing the manuscript and Rosemarie Shambrook for drafting the figures included in this book. We also wish to thank the lecturers and the participants in the 1981 ECMWF Seminar for contributing with manuscripts and useful ideas to this book.

The Editors, Reading and Stockholm, 1983

Contents

List of Authors	IX
List of Participants	XI
Some Aspects of Atmospheric Predictability E. LORENZ	1
Medium Range Forecasting at ECMWF; A Review and Comments on Recent Progress L. BENGTSSON	21
Current Problems in Medium Range Forecasting at ECMWF; Model Aspects A. SIMMONS	43
Current Problems in Medium Range Forecasting at ECMWF; Data Assimilation Scheme G.J. CATS	69
Predictability of Time Averages: Part I: Dynamical Predictability of Monthly Means	109
Part II: The Influence of the Boundary Forcings J. SHUKLA	155
Statistical Methods for the Verification of Long and Short Range Forecasts C. LEITH	207
Bifurcation Mechanisms and Atmospheric Blocking E. KÄLLÉN	229
Dynamically Stable Nonlinear Structures C. LEITH	265

List of Authors (Present at the Seminar)

A. GILCHRIST
Meteorological Office
London Road,
Bracknell, Berks.
England

C. LEITH
National Center for Atmospheric Research
P.O. Box 3000
Boulder, Colorado 80307
USA

E. LORENZ
Department of Meteorology
Massachusetts Institute of Technology
Cambridge, Massachusetts 02139
USA

R. MADDEN
National Center for Atmospheric Research
P.O. Box 3000
Boulder, Colorado 80307
USA

J. SHUKLA
Laboratory for Atmospheric Sciences
NASA/Goddard Space Flight Center
Greenbelt, Maryland 20771
USA

L. Bengtsson

G. Cats

U. Cubasch

P. Källberg

E. Källén

A. Simmons

S. Uppala
European Centre for Medium Range
Weather Forecasts, Shinfield Park, Reading, Berkshire RG2 9AX/England

List of Participants

S. Bodin
SMHI
Norrköping, Sweden

S. Brière
Météorologie Nationale
Paris, France

D.J. Carson
Meteorological Office
Bracknell, UK

D.R. Davies
Exeter University
Exeter, U.K.

A. van Delden
Utrecht University
Utrecht
The Netherlands

M. Déqué
Météorologie Nationale
Paris, France

St. Emeis
Universität Bonn
Bonn, FR Germany

P.J. Everson
Exeter University
Exeter, UK

C.K. Folland
Meteorological Office
Bracknell, UK

A.M. Gadian
Exeter University
Exeter, UK

H. Gmoser
Zentralanstalt für Meteorologie
u. Geodynamik Wien, Austria

J.S.A. Green
Imperial College
London, UK

M. Hantel
Universität Bonn
Bonn, FR Germany

K. Hedegaard
Meteorological Institute
Copenhagen, Denmark

B. Hoskins
Reading University
Reading, UK

Å. Johansson
Stockholm University
Stockholm, Sweden

D.E. Jones
Meteorological Office
Bracknell, UK

A.M. Jørgensen
Meteorological Institute
Copenhagen, Denmark

A. Kakouros
National Meteorological Service
Athens, Greece

G.P. Können
KNMI
De Bilt, The Netherlands

A. Lecluyse
Institut Royal Météorologique de Belgique
Bruxelles, Belgium

H. Lejenäs
Stockholm University
Stockholm, Sweden

C. Lemcke
KNMI
De Bilt, The Netherlands

J. Luo
WMO
Geneva, Switzerland

M.K. MacVean
Reading University
Reading, UK

R.H. Maryon
Meteorological Office
Bracknell, UK

L. Moen
SMHI
Norrköping, Sweden

E. Müller
Deutscher Wetterdienst
Offenbach, FR Germany

P. Nalpanis
Imperial College
London, UK

S. Nickovic
University of Belgrade
Belgrade, Yugoslavia

S. Nilsson
MVC
Stockholm, Sweden

E. Olofsson
Stockholm University
Stockholm, Sweden

E. Özsoy
Turkish State Meteorological Service
Ankara, Turkey

R.P. Pearce
Reading University
Reading, UK

M.A. Pedder
Reading University
Reading, UK

B. Rajkovic
University of Belgrade
Belgrade, Yugoslavia

C.E. Reeve
Exeter University
Exeter, UK

V. Renner
Deutscher Wetterdienst
Offenbach, FR Germany

R. Reynolds
Reading University
Reading, UK

J. Siméon
Météorologie Nationale
Paris, France

G.H. White
Reading University
Reading, UK

G. Wu
Imperial College
London, UK

Some Aspects of Atmospheric Predictability

E. LORENZ

1. PREDICTABILITY, DETERMINISM, STABILITY, AND PERIODICITY

1.1 Introduction

Our central concern in this set of lectures will be the predictability of the atmosphere. Many of our presently accepted ideas on this subject have been gained from studies of other systems, including complex mathematical models of the atmosphere and simple sets of mathematical equations, and we shall consider the predictability of these systems also.

By the predictability of a system we shall mean the degree of accuracy with which it is possible to predict the state of the system in the near and also the distant future. The predictability which we shall consider will be qualified rather than intrinsic, in that we shall assume from the outset that the predictions will be based upon less than perfect knowledge of the system's present and past states. This qualification would not be needed if we were interested only in mathematical models, but, for real physical systems, anything else would be unrealistic. For example, our present-day instruments may be capable of measuring winds to the nearest meter per second, but we cannot imagine that instruments in the near future will routinely measure them to the nearest centimeter per second. Likewise, our present global observing network may resolve the structure of the principal cyclones and anticyclones, but we cannot visualize observations in the near future which will regularly reveal the location, size, and internal structure of every cumulus cloud.

We wish to make our treatment of qualified predictability rather general. In this lecture we shall introduce the concepts of multiple time series and processes, and determinism and randomness. We shall then show that we lose virtually no generality by considering only the predictability of deterministic time series which are realizations of processes defined by mathematical equations.

1.2 Prediction and its accuracy

If we are to consider the accuracy with which it is possible to predict, we must first agree on what is meant by prediction. Consider the following hypothetical predictions for tomorrow's temperature at a particular city, say London.

1. Tomorrow will be warm (i.e., warmer than normal).
2. Tomorrow will be very warm.
3. Tomorrow's maximum temperature will be 27°C.
4. The probability that tomorrow will be warm is 80 per cent.

If we disregard the occasional cases when tomorrow will be warm on one side of the city and cool on the other, for example, or warm in the morning and cool in the afternoon, it is evident that the first prediction will be either right or wrong. It is equally evident that someone who knows nothing about weather forecasting will, if he simply predicts that the following day will be warm or cool, be right half of the time. One therefore cannot demonstrate skill in forecasting simply by making a few correct forecasts and then quitting; he must show that in the long run he is right more often than wrong.

If we agree on how warm is "very warm", the second prediction is also either right or wrong. However, since a forecaster can forecast "very warm" when it turns out to be only marginally warm, or vice versa, the second hypothetical prediction will be right less often than the first, if made by a forecaster of equal and less than perfect skill. It follows that any absolute measure of skill must take into account the manner in which the prediction is worded.

If we agree that the "maximum temperature" means the officially recorded maximum, rounded to the nearest whole degree, the third prediction also is either right or wrong, and, unless the forecaster is very skillful, it is probably wrong. Yet in an obvious sense the forecaster who predicts 27°C when the temperature turns out to be 26°C has made a better forecast than the forecaster who predicts 21°C or 31°C. For forecasts of this sort an obvious measure of accuracy is the magnitude of the error, or some function thereof. To serve as a measure of skill, this quantity should be averaged over many forecasts, since one might always guess well on a single occasion. The mean-square error, and also its square root, have become rather widely used measures of the accuracy of a sequence of numerically expressed forecasts, and we shall assume that they are adequate measures for our purposes, provided that the forecasts are uniformly expressed. If we wish to compare a forecast worded in one manner with one worded in another, we must also decide how large a mean-square error for one type of forecast is equivalent to a given mean-square error for the other.

The first prediction could also be made numerical, perhaps by letting "1" represent "warmer than normal", while "0" represents "cooler than normal". The forecaster who predicts 1 or 0 then makes a squared error of one unit whenever he is wrong, and zero whenever he is right, and his mean-square error is the fraction of the time that he is wrong.

The forecaster will soon recognize that he can obtain a smaller mean-square error by frequently predicting a fraction, even though the outcome must be 1 or 0. For example, if he knows nothing about forecasting, he can reduce his mean-square error from 0.5 to 0.25 by predicting 0.5 all of the time. If he recognizes a situation which will be followed by warm weather 80 per cent of the time, he can, whenever this situation occurs, predict 0.8, thus making a squared error of 0.04 in four fifths of these cases, and 0.64 in one fifth, and hence obtaining a mean-square error of 0.16 for these cases, while he would have obtained 0.2 if he had predicted 1.0 on each occasion. He should not be informed at this point that predicting a fraction is illegal, for he is actually making the fourth hypothetical prediction, i.e., stating the probability of warmer weather, which in some instances will be the most useful statement which he could make.

The point of this discussion, as far as predictability is concerned, is that whether one is forecasting categories or numerical values, and most likely outcomes or probabilities, the forecast may be expressed in numerical terms, and the square of the difference between the predicted number and the outcome will serve as a measure of the accuracy of a given forecast. We shall therefore lose no generality in assuming in these lectures that both the forecast and the outcome are always numbers. The skill of a forecaster, or of a forecasting system, in predicting a particular quantity on many occasions can then be measured in terms of the mean-square or root-mean-square error. Since the usual weather forecast involves predicting many weather elements, the mean-square errors in predicting the different elements must be weighted in some manner before being averaged. We must decide, for example, how many meters per second of wind are equivalent to one degree of temperature. In inquiring as to predictability we shall then be asking how small a suitably weighted mean square error can be made.

1.3   Time series and processes

The sequences of weather elements in whose predictability we are interested are examples of time series. By a simple time series we shall mean merely a function of time. It may be a function of continuously varying time, for example, the temperature at London, or it may be defined only for discrete values of time, for example, the daily maximum temperature at London. It may be defined in physical terms, or as a mathematical function.

A <u>multiple</u> <u>time</u> <u>series</u> is an array of simple time series. For the concept to be useful, the separate time series should be physically related. For example, they might be temperatures at neighboring cities or temperature and wind speed at the same city, or they might form a particular solution of a system of several coupled ordinary differential equations.

A <u>process</u> (simple or multiple) is an ensemble of time series; the separate members of the ensemble are <u>realizations</u> of the process. For the concept to be meaningful, the separate realizations must be similarly produced. For example, they may be separate time-dependent solutions of a single system of differential equations which possesses many solutions.

The global weather may be regarded as a single realization of a multiple process. We can visualize the process by imagining an ensemble of planets, each identical to the earth, and each possessing an atmosphere governed by the same laws, but with different global weather distributions at some key time. In view of our earlier discussion, we may restrict our attention to the predictability of time series, and we may assume that these series are realizations of readily definable processes.

1.4   Determinism and randomness

A process is <u>deterministic</u> if the present state of a realization completely determines the state at any specified future time, i.e., if two realizations which are identical at one time must be identical at all future times. A process is <u>random</u>, or <u>stochastic</u>, if the present state of a realization merely determines a probability distribution of states at a specified future time; two realizations may therefore be identical at present and differ in the future. If all possible present states determine the same future probability distribution at all future times, the process is <u>completely</u> <u>random</u>; in this case knowledge of the present state of a realizations tells us nothing about its future. So-called tables of random numbers are intended to be realizations of completely random processes. The weather is presumably a realization of a random process, since its behavior depends to a certain extent on human activity, which we hesitate to call deterministic. Most mathematical models of the atmosphere define deterministic processes.

Sometimes a process is considered to be deterministic if the present and past states of a realization determine the future, even though the present state alone may not completely determine the future. A process is then considered random if it is not deterministic in this sense. A process which is deterministic in this sense may be random in the former sense; an example of such a process would be one defined by a single second-order difference equation. We have chosen the former definition simply as a matter of convenience.

It might appear at this point that realizations of a deterministic process would be perfectly predictable, while those of a random process would be less than perfectly predictable, their predictability at any future time depending upon the variance of the probability distribution determined by the present state of the realization. From our point of view this is actually not the case, because of the qualification which we have imposed upon prediction, namely, that our knowledge of the present state of the realization must be imperfect.

To see how this qualification affects predictability, we note first that a deterministic process may be converted into a random process in either of two ways. First, we may round off the numerical values, thereby no longer distinguishing between states which are nearly identical. In all likelihood the set of states which are rounded off to a single state will be followed by a set of states which will be rounded off to two or more distinct states; the new process consisting of rounded-off numbers will therefore be random. Second, we may reduce the multiplicity of the process by discarding certain series from the array of series which forms each realization. For example, we may replace the weather, as it would behave if human activity and other random influences were somehow eliminated, by the weather, as revealed by actual measurements spaced a few hundred kilometers apart. These measurements resolve the cyclones and anticyclones, but omit most of the cumulus clouds, which exert considerable influence on the cyclones and anticyclones. The process with reduced multiplicity would therefore be random even if the complete weather behavior were deterministic. We should note that a random process produced in this manner is sometimes deterministic according to the second definition of determinism.

At this point we may inquire whether the converse of this result is true, i.e., given any random process, does there exist a deterministic process which may be converted into this random process by rounding off the numerical values, or by reducing the multiplicity of the time series? We suspect that it would be difficult to prove that the answer to this question is affirmative. Nevertheless, we feel that for practical purposes the answer is "yes", i.e., that any random process which we are likely to encounter behaves similarly to one which may be derived from a deterministic process. We shall therefore confine our attention to the predictability of processes which are deterministic, and therefore representable as solutions of deterministic systems of equations, with the understanding that in making our predictions we cannot determine the initial state with perfect accuracy or completeness. We can then state that the qualified predictability of the deterministic process depends upon the variance of the future probability distributions determined by present states in a derived random process.

1.5 Stability and instability

To pursue this matter further we require some more definitions. For our purposes we may define the difference between two realizations of the same process, at a given time, as simply the absolute value of the difference of the two numerical values, if the process is simple, or as the square root of a suitably weighted average of the squares of the differences, if the process is multiple. We shall then say that a realization of a deterministic process is <u>stable</u> if the difference between this realization and any other realization remains small throughout the future, whenever the difference is sufficiently small at the present. Otherwise the realization is <u>unstable</u>; in this case one can find realizations whose distance from the given realization is arbitrarily small at present, but exceeds some preassigned value in the ultimate future.

It therefore appears that the deciding factor in predictability is not determinism vs. randomness, but stability vs. instability. If a realization is unstable, other realizations which are close enough to it at present to be mistaken for it will eventually be found far away from it. Hence even a forecast made by perfect extrapolation, which might follow any one of these realizations, will in most instances eventually depart from the particular realization which is the outcome. If, on the other hand, the realization is stable, a good although not perfect forecast into the indefinite future may be made by making the initial error small enough.

It is to be emphasized that these conclusions are not restricted to forecasts which are made by attempting to integrate the governing equations, or some approximation to them. If there are many possible realizations, each of which eventually departs from each of the others, no forecast, whether it be dynamical or empirical, and objective or subjective, can approximate more than one of these realizations, and if a forecast should follow the correct realization it would be a mere coincidence.

In the following lecture we shall be concerned with the <u>rate</u> at which small initial errors amplify, as time progresses. This rate has been most frequently estimated from numerical studies with mathematical models of varying degrees of complexity. In the mean time, we can often determine whether instability is present or not by examining the past history of the realization.

We first assume that the system with which we are dealing is <u>compact</u>, i.e., that the number of different states, each removed from each of the others by a pre-specified amount, is limited, for any realization. This is certainly true when the system is any one of the commonly used mathematical models of the atmosphere; it is probably true when the system is the atmosphere itself, and reasonable weighting functions are chosen in defining the difference between states. It then follows that any realization will eventually acquire a state which is arbitrarily close to one which it has previously acquired. If the realization is stable, history will then approximately repeat itself, and the system will be observed to vary periodically. Equivalently, if it is observed to vary aperiodically, even though occasional temporary near-repetitions occur, the realization must be unstable (cf. Lorenz 1963, Charney <u>et al</u>., 1966).

Observations of the atmosphere indicate that it is not a periodically varying system. It does have periodic components, notably the annual and diurnal cycles and their overtones. However, when these and any other suspected periodic components are subtracted from the total signal, a large residual is still present. The periodic components are predictable at arbitrarily long range, by mere repetition. The residual is what the weather forecaster usually wishes to predict, and its lack of periodicity indicates that it is unpredictable in the sufficiently distant future.

2. THE GROWTH OF ERRORS: WHY WE CAN'T PREDICT
2.1 General considerations

Let the equations governing a multiple deterministic process be

$$d\underline{X}/dt = \underline{F}(\underline{X}) ,\qquad(1)$$

where t is time, and the components of $\underline{X}$ are $X_1,---,X_N$. Let $\underline{X}^*$ be a realization of the process. Let a second realization be given by

$$\underline{X} = \underline{X}^* + \underline{x} ,\qquad(2)$$

and assume that the variables X_i have been normalized so that a suitable measure of the difference between the realizations is

$$D = (\underline{x} \cdot \underline{x})^{1/2} .\qquad(3)$$

Consider a case where at some "initial" time t_o, D is small. Then, during such time as the "errors" x_i remain small, they are governed approximately by the linearized equation

$$d\underline{x}/dt = \underline{\underline{A}}\, \underline{x}, \qquad (4)$$

where the elements A_{ij} of the matrix $\underline{\underline{A}}$ are the partial derivatives $\partial F_i/\partial X_j$, evaluated from the solution $\underline{X}^*$. Eq. (4) determines whether, if D is initially small enough, D remains small for all time, i.e., the realization is stable, or D eventually grows quasi-exponentially, i.e., the realization is unstable.

In the special case where $\underline{X}^*$ is a steady solution, the elements of $\underline{\underline{A}}$ are constants, and the stability of the realization is determined by the eigenvalues of $\underline{\underline{A}}$. If at least one eigenvalue has a positive real part λ, the solution is unstable, and D ultimately behaves like $e^{\lambda t}$; otherwise the solution is stable.

In the general case the elements of $\underline{\underline{A}}$ vary with time, and the eigenvalues of $\underline{\underline{A}}$ at any particular time do not indicate the stability of the solution. To determine from (4) whether D increases we would have to solve (4) from the appropriate initial conditions. The most feasible method for solving (4), in some cases the only feasible one, is by stepwise numerical integration. This would entail first solving (1), also numerically, to obtain the elements of $\underline{\underline{A}}$ at successive time steps. At this point it would become simpler to solve (1) twice, with slightly different sets of initial conditions, evaluating D at intervals to see whether it is increasing. Such a procedure has become the standard method for investigating the predictability of atmospheric models, and systems of similar complexity or irregularity. There is no need in such a study to make the initial errors as large as typical observational errors, and it is often convenient to make them much smaller. Such investigations have come to be known as "predictability experiments".

In the unstable case, once D has become large, the linearizing assumptions leading to (4) are no longer valid, and (4) cannot be correct in any event, since it would imply that D would grow forever. If (1) is realistic enough to force each variable X_i to remain within bounds, D must also remain bounded. Ultimately D should oscillate about a value no greater than the difference between two randomly chosen states of the system governed by (1), although it may exceed this value for extended periods of time. If D fails to reach this value, there is some predictability even at very long range, and from what we have said earlier, the system presumably varies with a periodic component.

2.2 A sample predictability experiment

In order to demonstrate how a predictability experiment works, we shall choose a simple system, which is, in fact, the simplest nonlinear system which can be used for such an experiment. The governing equation is a single first-order quadratic difference equation, which may be written

$$Y_{n+1} = a Y_n - Y_n^2 , \qquad (5)$$

where a is a constant. If $0 \leq a \leq 4$, and $0 \leq Y_0 \leq a$, Eq. (5) generates a sequence $Y_0, Y_1, Y_2, ---$ with $0 \leq Y_n \leq a$ for all n. For any particular value of a, the set of all such sequences constitutes a deterministic process. We have previously studied this equation in considerable detail (Lorenz 1964), and this and similar equations have recently received much attention from mathematicians (e.g., Guckenheimer 1977, Collet and Eckmann 1980). No claim is made that this process is a model of the atmosphere.

In Table 1 the first column of "data" is a segment of a particular solution of (5), with $a = 3.75$ and $Y_0 = 1.5$. Each value has been rounded off to four decimal places before being used to compute the next value. There is no evidence of periodicity in the behavior.

In the next column a small error $y_0 = 0.001$ has been added to Y_0, changing it from 1.5 to 1.501. The difference between the columns is observed to grow rather irregularly as n increases. We have underlined the first item in this column differing from the corresponding item in the former column by at least 0.001, 0.01, 0.1, and 1.0; on the average about 7 or 8 steps are needed for the error to grow by a factor of 10. While the error y_n is small it is governed approximately by the linearized equation

$$y_{n+1} = (a - 2Y_n)y_n , \qquad (6)$$

but of course it never becomes larger than a, since both columns are bounded by 0 and a.

In the next column we have set Y_0 back to 1.5 but have raised a to 3.751, in order to simulate the effects of predicting with an imperfectly known governing equation or physical law. Here too we find that the error increases irregularly, at a comparable rate. It appears to make little difference whether the original error is in the initial conditions or the governing equations. One can argue that, if the governing equations are wrong, an error will immediately be introduced, and will then grow just as if it had been present initially.

TABLE 1. Numerical solutions Y_n of the difference equation $Y_{n+1} = a Y_n - Y_n^2$, for indicated values of a and Y_o. Computation of each value has been carried to indicated number of decimal places

n	a			
	3.7500	3.7500	3.7510	3.750
0	1.5000	1.5010	1.5000	1.500
1	3.3750	3.3757	3.3765	3.375
2	1.2656	1.2635	1.2645	1.266
3	3.1443	3.1417	3.1442	3.145
4	1.9045	1.9111	1.9079	1.903
5	3.5148	3.5143	3.5165	3.515
6	0.8267	0.8283	0.8246	0.826
7	2.4167	2.4200	2.4131	2.415
8	3.2222	3.2186	3.2285	3.224
9	1.7007	1.7104	1.6869	1.696
10	3.4852	3.4885	3.4819	3.484
11	0.9229	0.9122	0.9370	0.927
12	2.6091	2.5886	2.6367	2.617
13	2.9767	3.0064	2.9381	2.965
14	2.3019	2.2356	2.3884	2.328
15	3.3334	3.3856	3.2544	3.310
16	1.3887	1.2337	1.6161	1.456
17	3.2791	3.1044	3.4502	3.340
18	1.5441	2.0042	1.0378	1.369
19	3.4061	3.4989	2.8158	3.260
20	1.1714	0.8786	2.6333	1.597
21	3.0206	2.5228	2.9432	3.438
22	2.2032	3.0960	2.3775	1.073
23	3.4079	2.0248	3.2655	2.872
24	1.1658	3.4932	1.5854	2.522
25	3.0127	0.8971	3.4333	3.097
26	2.2213	2.5593	1.0908	2.022
27	3.3957	3.0474	2.9017	3.494
28	1.2031	2.1411	2.4644	0.894
29	3.0642	3.4448	3.1707	2.553
30	2.1014	1.0514	1.8400	3.056

In order to simulate the effects of being unable to devise a perfect mathematical procedure for solving a system of equations, even if we should know the equations perfectly, we have added a final column to Table 1 in which everything is rounded off to three instead of four places. Again the error increases irregularly, at nearly the same rate. By coincidence all four solutions are reasonably close again at step 27, but by step 30 they are noticeably farther apart again. Our general conclusion is that as long as the error is initially not too great, its origin is immaterial; the time required for it to become large will depend upon the typical rate at which separate solutions of the appropriate system of equations diverge from one another.

2.3 Experiments with atmospheric models

Long before predictability experiments were conceived, Thompson (1957) had deduced theoretically that small errors should tend to amplify. As far as we can determine, the first predictability experiments were made with highly simplified atmospheric models which attempted to capture some of the gross features of atmospheric behavior. In the first of such studies to be written up in detail (Lorenz 1965), the process was governed by a set of 28 coupled ordinary differential equations. These equations contained quadratic terms representing the advection of vorticity and potential temperature, linear terms representing mechanical and thermal damping, and constant terms representing thermal forcing. The equations were derived from the familiar two-layer quasi-geostrophic model. The 28 variables made possible the representation, in each layer, of a zonal flow with two north-south modes, and superposed waves of three different east-west wave lengths, each with two north-south modes and sine and cosine east-west phases.

A period of 64 days was simulated. The growth rate of small errors was found to fluctuate wildly with the "synoptic situation"; during some four-day periods there was virtually no error growth, while, during others, the errors increased tenfold, but, on the average, small errors in wind velocity or temperature doubled in about four days.

Let us see what a four-day doubling time would imply regarding practical weather forecasting. A typical observational error in temperature may be as low as 1°C; it is probably not much less. In eight days, such an error would grow to 4°C, which would usually be considered tolerable. Reasonably good forecasts a week in advance should therefore be possible. In twenty days, however, the error would grow to 32°C, which would presumably be intolerable. Forecasts three weeks in advance should therefore be impossible. Even allowing for the slowdown of error growth as the errors become larger, one-month or several-month forecasts would seem to be out of the question.

As these results became disseminated through the meteorological world, it was recognized that the conclusions were too important to be trusted to models composed of a few dozen ordinary differential equations. Predictability experiments were soon made with the few large global circulation models then in existence (Smagorinsky 1963, Mintz 1964, Leith 1965); with the facilities of the middle 1960's these represented a large expenditure of computer time. As might have been anticipated, the models were sufficiently dissimilar to one another for the predictability studies performed with them to give contradicting results. Leith's model indicated no growth of errors at all; Smagorinsky's indicated a 10-day doubling time, while Mintz's showed a 5-day doubling time. For various reasons Mintz's result came to be the most generally accepted one (see Charney et al. 1966). The prospects for forecasting a month in advance began to grow dim, while an explanation seemed to be needed for our failure at that time to make good forecasts a week ahead.

As larger global circulation models came into being, with greater horizontal and vertical resolution, and more accurate representation of some of the physical processes, predictability experiments continued (e.g., Smagorinsky 1969, Jastrow and Halem 1970, Williamson and Kasahara 1971). The experiments contained numerous variants; sometimes, for example, the initial errors were random, while at other times they were confined to specific scales of motion or specific geographical regions. Provided that they were small enough in amplitude, however, their spectrum or location had relatively little effect upon the growth rate. Gradually the generally accepted doubling time decreased to three days or less.

Since the zonally averaged flow in the middle-latitude troposphere tends to be baroclinically unstable, the question arises as to whether the instability of time-variable flow, which gives rise to error growth in the global circulation models, is also a manifestation of baroclinic instability. The typical growth rate of errors does appear to be comparable to the growth rate of perturbations superposed upon baroclinically unstable zonal flows. However, time-variable flows possessing migratory waves can also be barotropically unstable (Lorenz 1972). Predictability experiments with models of two-dimensional turbulence (e.g., Lilly 1969) have yielded growth rates comparable to those given by global circulation models. We suspect that either barotropic or baroclinic instability alone would be sufficient to bring about a three-day doubling time, and that actually both types of instability are ordinarily present.

By and large the results of predictability experiments seem to have been more applicable to middle and higher latitudes than to the tropics. The physical processes in the tropics which lead to instability, or perhaps stability, and in particular those involving water in the atmosphere, have not always been well modeled. Moreover, variances tend to be small in the tropics, and, unless some special weighting function is introduced, the behavior of errors in the tropics will tend to be excluded from the behavior of some over-all mean-square error.

2.4 Evidence from observations

We could have greater confidence in our estimate of the doubling time if we could perform predictability studies with real observational data instead of numerical models. Unfortunately for this purpose, the atmosphere is not a controlled experiment; we can perhaps introduce some disturbances and see what happens, but then we shall never know what would have happened if we had not introduced the disturbances. We could demonstrate that predictability exists at a certain range by showing that a particular forecasting procedure consistently gives good forecasts, but, at least at short range, the numerical studies which indicate that predictability is limited nevertheless acknowledge the possibility of better forecasts than our present schemes produce.

The closest attainable approximation to a controlled experiment seems to be a study based on analogues. If within the historical records of past weather we can find two global weather patterns which are nearly identical, one pattern may be regarded as equal to the other pattern plus a small error, and, by examining what happened following the occurrences of the two patterns, we can see how rapidly the error grows.

Unfortunately, in the only study of this sort of which we are aware (Lorenz 1969a), based upon five years of data, we found only two pair of patterns whose differences were as small as 62 per cent of the difference between randomly chosen patterns; we estimated that 130 years of data would be needed to reduce this difference to 50 per cent. We were forced, therefore, to study the growth of errors which were already rather large, and which would never double again.

We found that, on the average, the smaller errors tended to amplify by about 9 per cent in one day, which would correspond to an eight-day doubling time if this amplification rate could continue. The most reasonable extrapolation of the results to very small errors, however, indicated that the latter would double in about 2.5 days - a result which is gratifyingly close to the doubling times suggested by the numerical models.

It would be possible, of course, to find weather patterns in the historical data which closely resemble one another over limited areas. There is no reason, however, to believe that these patterns will evolve similarly if the patterns in the surrounding areas are not similar. The results of a limited-area analogue study would therefore be hard to interpret.

2.5 The influence of smaller scales

The errors which have been estimated to double in about three days are of necessity the errors in the larger scales of motion. Features with horizontal scales of less than a few hundred kilometers are not resolved by the grids in the global circulation models, nor by the network of observing stations. Nothing in these studies indicates how rapidly the errors in the smaller scales would grow, if these scales were resolved by the models or the map analyses. It seems likely, however, that they would grow much faster; errors in describing the structure of a thunderstorm, for example, should grow as rapidly as the thunderstorm itself, doubling in perhaps half an hour or less.

It is beyond the capability of any computer to handle a global model which resolves individual thunderstorms and features of similar size. Accordingly, we derived a system of equations whose dependent variables were the variances of the errors in velocity in successive octaves of the horizontal spectrum (Lorenz 1969b). The coefficients in these equations depended upon the variance spectrum of the velocity itself, which we specified in advance. We found that errors in the larger scales doubled in a matter of days, while those in the smaller scales doubled in hours or minutes. More importantly, however, the errors in any scale, once established, soon induced errors in adjacent scales. Thus, even if there were no initial errors in the larger scales, the initial errors in the smallest scales would progress rapidly to slightly larger scales and thence more slowly to still larger scales, so that, after a day or so, there would be appreciable errors in scales of 1000 kilometers or greater. These would then proceed to double in a few days, just as if they had been present initially.

The model which yielded these results was much too crude to be accepted uncritically. It was derived from the barotropic vorticity equation governing flow over a homogeneous plane region, rather than the complete atmospheric equations over a sphere with tropics and polar regions, oceans and continents, and mountains and plains. There were no baroclinic effects nor effects of moisture, and no forcing nor damping. Finally, it was necessary to introduce a closure approximation of questionable validity.

Subsequent studies, however, using more sophisticated closure assumptions
(e.g., Leith 1971, Leith and Kraichnan 1972), have yielded qualitatively similar
results. Robinson (1967), on the other hand, had arrived at similar conclusions
from the premise that the equations usually considered to govern the atmosphere
were actually inapplicable to an atmosphere with a continuous spectrum of motions.

What then, do we mean when we speak of a three-day doubling time? We could
mean the rate at which errors in the larger scales would double if there were
no superposed smaller scales, or if the statistics of the smaller scales were
exactly determined by the larger scales which they accompanied. Alternatively,
we could mean the rate at which the large-scale errors will grow once they have
attained a size somewhat smaller than present-day observational errors, which,
in view of the smaller-scale errors, will be no more than a day or so after a
hypothetical time when the large scales might be error-free.

If the smaller scales were completely absent, we could extend the range of
acceptable forecasting by three days simply by cutting the observational errors
in half; this we could conceivably do two or three times, thereby adding a
week or so to the range of practical predictability. As things stand, however,
we can gain something by improving our observations, but we must eventually
reach a point where the observational errors will be no larger than the errors
which would be present anyway after a day or so, because of the smaller scales.
After that point is reached, further improvements in observations will yeild
only minor improvements in medium and long range prediction.

3. REGIMES AND EXTERNAL INFLUENCES: WHY WE CAN PREDICT
3.1 General considerations

The results discussed in the previous lectures suggest that further progress
can be made in short-range forecasting, but they are quite pessimistic for
medium-range and especially long-range forecasting. Such complete pessimism
may not be warranted.

In many mathematical models of the atmosphere, the only important nonlinear
terms are those representing advection. These terms are quadratic; they contain
the product of the advected quantity with the wind which does the advecting.
These are the terms which are responsible for the growth of small errors. They
are well formulated in the models.

Once the errors have become moderately large, the processes which are responsible
for slowing and eventually stopping their growth may not include advection.
Whatever these processes are, they need not be well represented in a model
which represents advection properly.

One possible manner of behavior is that the errors, as they become large, may rapidly approach their limiting magnitude. On the other hand, they may temporarily cease or nearly cease to grow long before they reach this magnitude. Still another possibility is that the errors, although large, may rapidly oscillate in sign, so that averages over periods of a few days or longer may be predictable even when instantaneous states are not. This would be the case, for example, if we could predict the passage of a sequence of storms over a period of days, without being able to say on which days during the period a storm would cross a given longitude. It is possibilities of the latter sort which offer hope for medium-range and long-range forecasting, and perhaps for predicting changes in climate. Since the cause of behavior of the latter sort, if it exists, is uncertain, the ideas presented in this lecture are necessarily more speculative than those in the previous ones.

3.2 Slowly varying features

There are two factors which might cause these possibilities to be realities. One is the existence of features which by their physical nature tend to vary slowly. These may be external to the atmosphere, but they exert an influence upon the atmosphere.

Probably the most frequently mentioned feature of this sort is the ocean-surface temperature field. This tends to vary slowly because of the ocean's large heat capacity. It also exerts a direct thermal influence upon the atmospheric temperature field. The atmosphere in turn exerts a direct thermal influence upon the ocean temperature, but it cannot alter it so rapidly. Thus, for example, if a certain area of the ocean is observed to be warmer than normal, there may be a physical basis for predicting that the atmospheric temperature a few weeks in advance, in the area influenced by this part of the ocean, will be somewhat warmer than normal rather than simply normal. Studies which have compared observed oceanic temperatures with subsequent atmospheric temperatures have not all agreed. Davis (1976) and Haworth (1978), for example, obtained somewhat nevative results, while Shukla and Misra (1975), and, in a later study, Davis (1978), obtained positive results. It appears possible that the phenomenon is rather important in the tropics but less so in higher latitudes.

Other features of this sort include the abundance of sea ice and continental snow cover. Qualitatively, sea ice and snow cover would seem to have the same effect as abnormally low ocean temperature, but the physical mechanism involved is different, since, in addition to directly cooling the atmosphere, ice and snow reflect much of the solar radiation which would otherwise warm the atmosphere.

Still another feature is the variability of solar radiation. In particular, solar anomalies associated with the sunspot cycle vary rather slowly. The general proposition that the atmosphere should respond to variations in solar activity is entirely reasonable, although the precise nature of the expected response is hard to deduce. Here also, observational studies have been suggestive but inconclusive.

3.3 Atmospheric regimes

The other factor favoring medium-range and long-range predictability is the existence of what might be called atmospheric regimes. During a regime the atmosphere tends to behave in one particular anomalous manner; when the regime changes, the atmosphere behaves for a considerable time in a different anomalous manner. The extreme case of regime behavior is a phenomenon called "almost intransitivity". This exists if there are two or more sets of possible weather patterns, and, as a result of its internal dynamics, the atmosphere can evolve readily from one pattern to another one in the same set, but only with difficulty to a pattern in another set. Almost intransitivity is easily produced in some of the simpler atmospheric models, but it is not certain that it occurs in the more realistic global circulation models, and its presence in the real atmosphere has not been verified observationally.

One of the most frequently mentioned regime phenomena is "blocking". This is characterized by the continued presence of troughs and ridges at preferred longitudes over extended periods, even though the long-term average circulation may not possess such troughs and ridges. Blocking has at times been called an illusion, but present evidence (Dole 1981) indicates that a blocking anomaly, once established, has a greater-than normal probability of persisting over any fixed time interval. Forecasting that blocking will persist is therefore somewhat better than guesswork, i.e., some extended-range predictability exists in such situations.

Another phenomenon, which is better documented than blocking, is the quasi-biennial oscillation, which is an oscillation between persistent easterly and persistent westerly winds in the equatorial stratosphere, generally taking somewhat over two years to complete a cycle. Here there is no question but what one can forecast a year in advance, and do much better than guesswork, simply by forecasting the upper winds to be westerly if they are presently easterly, and vice versa.

More generally, there appears to be on the average a slight tendency for local positive or negative anomalies to persist; this persistence does not completely die out in fifteen days (Lorenz 1973). Whatever the cause of the

persistence may be, its presence indicates that there is some predictability fifteen days in advance. Likewise, there is evidence that anomalies in averages over a given period persist over a considerably longer period; seasonal averages, for example, show some persistence a year ahead (cf. Madden 1977).
Possibly we should not have considered the regime phenomena separately from the external influences. Some investigators would maintain that regimes are externally caused.

3.4 Beyond persistence

What are the prospects for extended-range forecasts superior to those based upon pure persistence? One phenomenon which suggests that they are favorable is the Southern Oscillation, which is a dominating feature of both atmospheric and oceanic behavior in the tropical Pacific area. This is characterized by a set of events whose beginning may be hard to predict, but which, once established, seems to proceed in a rather predictable sequence (see Bjerknes 1969, Horel and Wallace 1981).

As for forecasting procedures, if we can identify the factors which are responsible for the cessation of the growth of larger errors, we can incorporate them into our models, just as we now include advection, which is responsible for the growth of small errors. In any event, we can immediately include some of the suspected factors. We can use models in which the ocean surface temperature field and the fields of sea ice and continental snow cover are variables. We can experiment with models where the solar output is variable, to see whether they behave differently from those where it is constant. Other candidates for incorporation or more refined formulation in future models appear to include various phenomena associated with water in the atmosphere, including clouds and precipitation, and the effects of these upon short-wave and long-wave radiation. These may be particularly important in establishing our ideas regarding the predictability of tropical weather.

4. REFERENCES

Bjerknes, J. 1969 Atmospheric teleconnections from the equatorial Pacific. Mon. Wea. Rev., 97, 163-172.

Charney, J. G. et al. 1966 The feasibility of a global observation and analysis experiment. Bull. Amer. Meteor. Soc., 47, 200-220.

Collet, P., and J.-P. Eckmann 1980 On the abundance of aperiodic behavior for maps on the interval. Commun. Math. Phys., 73, 115-160.

Davis, R. E. 1976 Predictability of sea-surface temperature and sea level pressure anomalies over the Northern Pacific Ocean. J. Phys. Oceanog., 6, 249-266.

Davis, R. E. 1978 Predictability of Sea Level Pressure Anomalies over the North Pacific Ocean. J. Phys. Oceanog., 8, 233-246.

Dole, R. M. 1981 Persistent anomalies of the extratropical Northern Hemisphere wintertime circulation. Ph.D. thesis, Mass. Inst. Technology.

Guckenheimer, J. 1977 On the bifurcation of maps of the interval. Inventiones Math., 39, 165-178.

Haworth, C. 1978 Some relationships between sea surface temperature anomalies and surface pressure anomalies. Quart. J. Roy. Meteor. Soc., 104, 131-146.

Horel, J. D., and J. M. Wallace 1981 Planetary-scale atmospheric phenomena associated with the Southern Oscillation. Mon. Wea. Rev., 109, 813-829.

Jastrow, R., and M. Halem 1970 Simulation studies related to GARP. Bull. Amer. Meteor. Soc., 51, 490-513.

Leith, C. E. 1965 Numerical simulation of the Earth's atmosphere. Methods in computational physics, Vol. 4. New York, Academic Press, 1-28.

Leith, C. E. 1971 Atmospheric predictability and two-dimensional turbulence. J. Atmos. Sci., 28, 148-161.

Leith, C. E., and R. H. Kraichnan 1972 Predictability of turbulent flow. J. Atmos. Sci., 29, 1041-1058.

Lilly, D. K. 1969 Numerical simulation of two-dimensional turbulence. Phys. Fluids, Suppl. II, 24-249.

Lorenz, E. N. 1963 Deterministic nonperiodic flow. J. Atmos. Sci., 20, 130-141.

Lorenz, E. N. 1964 The problem of deducing the climate from the governing equations. Tellus, 16, 1-11.

Lorenz, E. N. 1965 A study of the predictability of a 28-variable atmospheric model. Tellus, 17, 321-333.

Lorenz, E. N. 1969a Atmospheric predictability as revealed by naturally occurring analogues. J. Atmos. Sci., 26, 636-646.

Lorenz, E. N., 1969b The predictability of a flow which possesses many scales of motion. Tellus, 21, 289-307.

Lorenz, E. N. 1972 Barotropic instability of Rossby wave motion. J. Atmos. Sci., 29, 258-264.

Lorenz, E. N. 1973 On the existence of extended range predictability. J. Appl. Meteor., 12, 543-546.

Madden, R. A. 1977 Estimates of the autocorrelations and spectra of seasonal mean temperatures over North America. Mon. Wea. Rev., 105, 9-18.

Mintz, Y. 1964 Very long-term global integrations of the primitive equations of atmospheric motion. WMO-IUGG symposium on research and development aspects of long-range forecasting. World Meteor. Org., Tech. Note No. 66, 141-155.

Robinson, G. D. 1967 Some current projects for global meteorological observation and experiment. Quart. J. Roy. Meteor. Soc., 43, 409-418.

Shukla, J., and B. J. Misra 1975 Relationships between sea surface temperature and wind speed over the central Arabian Sea, and monsoon rainfall over India. Mon. Wea. Rev., 105, 998-1002.

Smagorinsky, J. 1963 General circulation experiments with the primitive equations. I. The basic experiment. Mon. Wea. Rev., 91, 99-164.

Smagorinsky, J. 1969 Problems and promises of determistic extended range forecasting. Bull. Amer. Meteor. Soc., 50, 286-311.

Thompson, P. D. 1957 Uncertainty of initial state as a factor in the predictability of large scale atmospheric flow patterns. Tellus, 9, 275-295.

Williamson, D. L., and A. Kasahara 1971 Adaptation of meteorological variables forced by updating. J. Atmos. Sci., 28, 1313-1324.

Medium Range Forecasting at ECMWF; A Review and Comments on Recent Progress

L. BENGTSSON

1. INTRODUCTION

The first ECMWF Seminar in 1975 (ECMWF, 1975) considered the scientific foundation of medium range weather forecasts. It may be of interest as a part of this lecture, to review some of the ideas and opinions expressed during this seminar.

It was generally recognised, and still is, that the mediumm range forecasting problem, defined as predictions within a time scale from 2 days to 2 weeks, was both a data- and a model problem. For this reason substantial efforts and resources were to be allocated to global analysis and initialization in addition to the modelling work. There was also a consensus of opinions that "the brute force approach" namely to use a high resolution general circulation type of model, was the only feasible method, in particular with respect to the Centre's time schedule and operational objective.

The practical experience in medium range weather prediction was very limited at this time and only a limited set of real data predictions of 2 weeks length had at that time been carried out at GFDL, Princeton (Miyakoda et al, 1972). As we will see later, these forecasts indicated in the average positive skill beyond a week, although the level of useful skill was small after 3-4 days. Nevertheless, these integrations represent a remarkable achievement and served as an important catalytic function. They stimulated the planning and later the successful implementation of the First GARP Global Experiment in 1978 - 79 and they constituted a very important impetus in setting up the ECMWF. They also set the benchmark for future efforts in medium range weather forecasting.

2. PROBLEM AREAS IN NWP IN 1974/75

There were a large number of obstacles to medium range weather prediction in 1975. Some of them still remain but some have been overcome at least partially. Table 1 lists a selection of the most serious problems as seen 6 years ago. They are divided into five different areas.

The data assimilation (analysis and initialization) has been discussed extensively by Bengtsson (1975) and we will here only highlight a few facts. The large increase of non-synoptic quantitative observations, mainly from satellites, which started in the 1970's, necessitated a continuous assimilation or alternatively a discontinuous assimilation with high temporal resolution (at least 6 hours). However, if this was going to be possible, ways were to be found to suppress the high frequency noise created by lack of a proper initialization before the next set of observations were inserted into the model. If this was not possible the noise level would successively increase to a level where new observations could not be properly assimilated any longer. The non-linear normal mode initialization developed simultaneously by Machenhaur (1977) and Baer (1977) presented a practical and a dynamically consistent solution to this problem at least at middle and high latitudes. Problems still remain in the Tropics where it is necessary to incorporate the effect of non-adiabatic processes. Recent work Wergen (pers.comm) has demonstrated that these processes can be included as well within the framework of the non-linear normal mode initialization.

Non-conventional data such as winds from aircraft and satellites need to be interpolated in the vertical as well as in the horizontal. In order to do this in a systematic way, a 3-dimensional analysis system based on optimum interpolation of observed increments was developed, Lorenc (1981).

Although this method has its limitations in rapidly changing situations (structure functions are statistical) it provides nevertheless an efficient

Table 1. Problem areas in medium range weather prediction in 1974/1975

Analysis	. analysis methods in the Tropics
	. usage of non-synoptic data
	. usage of non-conventional data
	. automatic control of observations in data sparse areas.
Initialization	. initialization in the Tropics
	. initialization of the ultra-long waves
	. initialization in the presence of mountains
	. the "spin-up" problem
Numerical methods	. computational stability in long term integration
	. grid representation vs spectral representation
	. grid representation in polar regions
	. semi-implicit algorithms
	. computational economy
Parameterization	. turbulent fluxes
	. radiation and clouds
	. heat balance
Surface representation	. orography and roughness
	. soil moisture and surface heat balance
	. air-sea interaction and representation of the oceans
	. surface albedo

method to analyse observations which are irregularly distributed in space and have different but presumably known error structure. The method is also very useful in automatic control of observations which is carried out by the same technique. Although simple methods are being applied to estimate the accuracy of the first guess and hence determine the relative weight given to a datum, further work is required before a satisfactory solution can be found.

The computational stability of the numerical model is provided by a finite different scheme which conserves potential enstrophy for a non-divergent flow. It uses a staggered grid of variables known as the C-grid (Arakawa and Lamb, 1977). This grid was selected because of its low computational noise and the ease of implementation of a semi-implicit scheme. It has excellent stability properties in extended integration as has been demonstrated by Sadourny (1975). Further details of the scheme can be found in Burridge and Haseler (1977), and Burridge (1979).

In spite of the very rapid development of spectral models around 1975 following the proposal put forward by Eliasen et al (1970) and Orszag (1970), to use an efficient transform technique, there was not enough practical experience in 1975 to select a spectral model. For this reason the Centre decided to use a grid point model as its first operational model. Since then, several meteorological services have implemented spectral models for operational numerical weather prediction. A comprehensive study Girard and Jarraud (1982) has demonstrated that a spectral model, using a triangular truncation, is superior to the operational grid point model at ECMWF for a comparative computational cost. For this reason ECMWF replaced the operational grid point model with a spectral one at the beginning of 1983.

A problem which was carefully considered 6 years ago, was the level of complexity or sophistication of the parameterization of sub-grid scale processes. Miyakoda and Sirutis (1977) evaluated 3 different levels of parameterization characterized by an increasing complexity of the description of vertical turbulent fluxes and convection. These results indicated an advantage in using the so-called higher order closure schemes (Mellor and Yamada 1974), while one of the most elaborate schemes to describe convection (Arakawa and Schubert, 1979) did not show the same advantage. ECMWF selected after some evaluation a scheme of moderate complexity. The parameterization of surface and turbulence fluxes was based on Monin-Obukov similarity theory and K-theory (first order closure) respectively, while Kuo's scheme (Kuo, 1974) was selected to describe convection. Louis (1979) has shown that this parameterization performed well in cases of intense air-sea interaction. The scheme also simulated in satisfactory details the structure of the boundary layer from the O'Neill experiment (Lettau and Davidson, 1957) and other similar experiments.

The radiative processes have been incorporated in great detail and the scheme is fully interactive with the large scale cloud-distribution (given as an empirical function of relative humidity and pressure). This scheme has been found to provide positive skill (Geleyn, 1980), as compared with a non-interactive cloud/radiation model (clouds geographically prescribed).

The physical processes at the surface and in the soil are described in a very simple heuristic manner. In 1975 a detailed specification of the surface condition was not believed to be essential for medium range forecasting although studies by e.g. Walker and Rowntree (1977) demonstrated the importance of soil moisture on the simulation of precipitation in the Tropics. The reason for this is the strong feedback between soil moisture and convective precipitation. This process is also of primary importance at higher latitudes in the summer, Rowntree and Bolton (1978), van Maanen (pers.comm.). In particular it takes a considerable time for the soil moisture and hence precipitation to adjust from an initial very dry to a normal or wet situation. During this time, which may take several weeks at middle and high latitudes and significantly longer in the Tropics or sub-tropics, precipitation is significantly reduced and there are associated changes on the dynamics due to change in diabatic forcing. For a more comprehensive discussion see Mintz (1983).

Due to the strong influence of the initial specification of soil moisture or the potential evapotranspiration it is absolutely essential to know it initially and it may not be an overstatement to claim that for medium range prediction it is probably more important to know the soil water or the potential evapotranspiration than to know the water vapour in the atmosphere!

3. OPERATIONAL ASPECTS

To establish an operational forecasting system implies a number of practical/technical constraints which must be considered carefully. Table 2 summarises some criteria which are crucial for model selection and model development and modification.

From the operational point of view the ECMWF plans called for an operational forecast once a day, which has to be produced in less than 8 hours including all operational aspects (decoding, analyses prediction, post-processing and dissemination). Of this time a 10-day prediction can occupy about 5 hours which means that the prediction must progress about 50 times faster than the evolution of the real weather. Given the constraints of the CRAY-1 computer this limited the resolution of the model to about 200 km and 15 vertical levels. The vertical resolution was selected because of the necessity to describe accurately enough the vertical structure of developing cyclones and the boundary layer. The horizontal resolution is still unsatisfactory and theoretical and synoptic investigations suggest that a resolution of around 100 km would be more satisfactory.

Table 2.   Model selection and model modificiation criteria

Practical/ Operational	. timing (50/1); 1 forecast a day
	. robustness (/failure/month)
Model formulation	. generality in formulation
	. lucidity (simplicity)
	. minimum of tuning
Programming aspects	. efficiency
	. flexibility toward future changes
Quality aspects	. objective evaluation
	. subjective evaluation

The model has to be designed in a way that calculations under no circumstances would create computational instability. Furthermore the reliability of the computer as well as the overall program structure must be such as to guarantee a very stable operational performance with less than 1 failure/month. This condition has been satisfied by a wide margin and the number of operational failures has been 5 since the start of operational forecasts in August 1979 and none at all in the operational year 1981 and 1982.

A very important factor in designing and maintaining forecasting systems of the complexity of that of the Centre is a clear and well defined structure both in the physical formulation and in the programming design.

From the very beginning the Centre has strived to maintain a generality in the physical formulation and to keep "tuning" to a minimum. Modifications of the model must therefore on the first hand be justified on dynamical and physical grounds and not only by the fact that a change happens to give better results. The reason is simply that it is extremely difficult to prove a better result on a limited set of numerical experiments only. Furthermore, conceptually simpler formulations have been selected in favour of more complex ones unless obvious improvements can be demonstrated.

The programming aspects have been constrained by computational efficiency and programming flexibility. While the first of these conditions is obvious, the second may need some justification. The Centre's forecasting system (data-asimilation and prediction model) contains the order of 100 000 lines of code. The additional support system such as pre- and post processing as well as verification and diagnostic evaluation is of a similar size. It is obvious that well-organized and well-documented programs must be set up if such a system would be fully useful. There are examples of large program systems of this kind which are impossible to use efficiently because the complexity is beyond a clear understanding - one has passed over the "threshold of complexity".

The careful design in programming implemented from the beginning at ECMWF has paid off and so far about 50 changes have taken place in the operational forecasting system during 3 years of operation. These changes often associated with major model and system changes, have been straightforward to implement and the number of support progammers have been very few.

4. IMPROVEMENT OF NUMERICAL FORECASTS

A very substantial improvement has taken place in numerical weather prediction since the very first forecasts were made more than 30 years ago. These improvements are essentially due to much better and more realistic models but also to the considerable increase in meteorological observations which have taken place over the period. This has been demonstrated by observing system experiments using FGGE data, e.g. Bengtsson (1983), where in particular observations from satellites and from aircraft have had a considerable impact on forecasts in the medium range (2-14 days). Furthermore, observations are now better utilised due to improved analysis methods, more accurate initialisation and a more accurate and consistent use of the prediction model to provide the first guess. Nevertheless, as can be seen below model development has had a fundamental effect on predictive skill. Results of 24 operational and quasi-operational 24 h predictions by the barotropic model (Staff Members, University of Stockholm, 1954) show an average standard deviation error of 76 m when verified over an area covering Northwestern Europe and Northeastern Atlantic. The integrations were done over a limited area (5700 km)^2 extending about 1200 km beyond the area of verification.

In order to compare the performance of the barotropic model with the Centre's operational model, we have carried through daily barotropic forecasts for the month of Janury 1981. The horizontal resolution was 38/km. In this case the integration domain was almost hemispheric and we used the initial state produced operationally. It is to be expected that these initial states are more accurate than those of 30 years ago, due to a more systematic use of observations at other levels and to a better first guess provided with the ECMWF model. The importance of an accurate initial state was in fact demonstrated, Staff Members, University of Stockholm (1954), and better forecasts were obtained when more carefully analysed data were used. Table 3 shows the standard deviation error for the barotropic model of 1954 and 1981 as well as the result of the operational ECMWF model. The verification for the latter experiment has been carried out over a domain covering Europe, 72N - 36N and 12W - 42E. This verification area is somewhat larger and in a more easterly position than the one used in 1954. The figures in the table demonstrates the achievements in numerical weather prediction over the last 30 years. It is also of interest to compare the computer time for a 24 hour forecast. The Staff Members, University of Stockholm were using the Swedish constructed BESK computer which about this time was one of the fastest if not the fastest computer in the world. BESK had an internal William-type memory; 512 words of 40 bits. No external memory was available at this time. Table 4 shows the necessary computer time (elapsed time) to produce a 1 day forecast. It should be pointed out that while both the BESK-code and the operational code on CRAY-1 are optimised, the barotropic code on CRAY-1 is non-vectorized standard Fortran.

Another example of the improvement in numerical weather prediction can be found from a recent study by Bengtsson and Lange (1982). In this study operational 3-day forecasts from different centres are compared. It is found that the forecasts by high resolution ECMWF model for instance have errors which are less than 60% compared to those based on less advanced filtered models.

Table 3. Standard deviation error for 500 mb forecasts over Europe

Time	Jan. 81 (31 cases)		Nov. 51 - April 54 (24 cases)
	ECMWF Operational model	Barotropic model	Barotropic model
24 h	22 m	47 m	76 m
48 h	41 m	97 m	
72 h	62 m	151 m	

Table 4. Computer time for a 24 h forecast

CRAY-1 ECMWF model (15 levels)	CRAY-1 Barotropic model	BESK Barotropic model
18500 grid points	2350 grid points	400 grid points
1100 sec	0.5 sec [1]	2400 sec [2]

[1] This time could be reduced to about 0.1 sec by redesigning the code
[2] Machine code

We will next study the improvement of medium range prediction at ECMWF since operational forecasts started in August 1979. An operational model normally goes through a process of improvements due to refinements in the treatment of physical parameterization and due to identification and correction of erroneous formulations and program errors. Such improvements have taken place both in the model and in the data-assimilation and are successively inserted into the operational forecasting system. Figure 1 shows the anomaly correlation for the 500 mb prediction for the winter of 1979/80 and 1980/81. It can be clearly seen that the scores for the second winter are significantly higher, reflecting a genuine improvement of the model. We have also incorporated as a benchmark for this comparison the very early results by Miyakoda et al, 1972, which constitute the first comprehensive trial of medium range predictions. Their ensemble mean anomaly correlation for the extra-tropical Northern Hemisphere was based on 12 January cases taken from the years 1964 to 1969.

5. POTENTIAL IMPROVEMENTS OF MEDIUM RANGE FORECASTS

The possibility of a further improvement of the numerical forecasts and the possibility of extending the length of useful forecasts is a fundamental issue of this seminar. As Professor Lorenz has pointed out in his lecture, there is an inherent limitation in the predictability of the atmosphere which is possibly of the order of a few weeks. However, there are also clear indications, Lorenz (1982) by comparing the internal error growth of the ECMWF model, with forecast error growth that there is scope for extending the range of useful forecasts by 3 - 4 days even with today's incomplete and inaccurate observations. These improvements are likely to fall into the largest scale of motion where the predictions have considerable errors of a systematic nature, Bengtsson and Simmons (1983). We will illustrate the

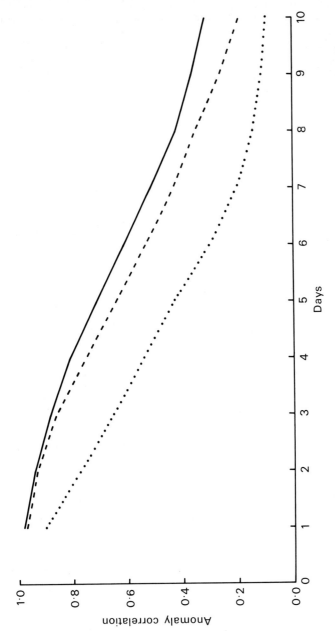

Fig. 1. Anomaly correlation for the 500 mb flow (Northern Hemisphere extra-tropics) for the winter 1980/81 (Dec., Jan., Feb.), full lines, for the winter 1979/80, dashed lines and Miyakoda et al. (1972) dotted lines

characteristic features of the systematic error by simply correcting the forecast from January 1981 by the monthly mean error of January 1980. The result is shown in Table 5. The table shows the extension of useful predictability in days as defined by 2 different levels of anomaly correlation. It is interesting to note the particular improvement for the ultralong waves, although the whole spectrum has in fact improved.

These experiments strongly suggest that there are still fundamental errors in the maintenance of the large scale quasi-stationary flow and the "climate" of the model differs from the real climate. A careful evaluation of the error, not reported here, shows that the predicted "quasi-stationary" flow has a smaller amplitude than the observed. Furthermore as can be seen from Fig. 2 which shows the observed 500 mb geopotential for January 1981 and the averages of all the 10-day forecasts for January 1981 the predicted large scale features are in a more easterly position.

It is suggested that a cause for this error is due to a reduced poleward transport of heat and momentum by the quasi-stationary flow which in turn is compensated by an increased transportation undertaken by the transient eddies which will develop into more active disturbances. A particularly interesting aspect of the model, supporting this suggestion is illustrated by Figure 3, where we study the performance of the operational model with respect to the energy cycle and to the kinetic and availabe potential energy budget. The figure shows the global energy budget for January 1981 as calculated from 12Z analyses as well as the ensembles of 2, 5 and 10-day forecasts. All standard levels between 1000 and 30 mb have been used. There is an overall increase of around 10% in total available and in zonal kinetic energy. The increase of the zonal part is greatest during the last five days. The eddy kinetic energy on the other hand is decreasing by a similar percentage. The energy cycle as measured by the dissipation rate is successively intensifying from an initial value of 1.7 Watts/m^2 towards a value of 3.3 Watts/m^2. The

Table 5. Predictability in days for 500 mb for January 1981 for the ECMWF model with and without statistical correction (for further information see text). Verification $20°N - 82.5°N$

	Anomaly correlation score (A.C.)	Number of days required to reach A.C. value	
		Before	After
		Statistical correction	
Ultra long waves (1 - 3)	80% 50%	4.3 8.9	4.8 >10
Medium waves (4 - 9)	80% 50%	3.5 6.0	3.7 6.4
Short waves (10 - 20)	80% 50%	1.4 3.2	1.5 3.3
Total field (all waves)	80% 50%	3.9 6.8	4.3 8.3

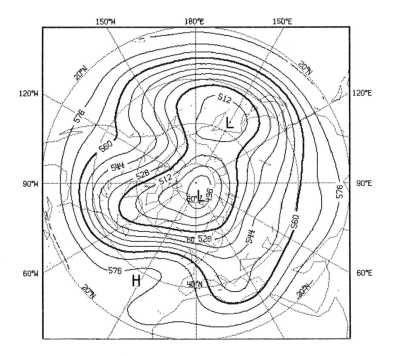
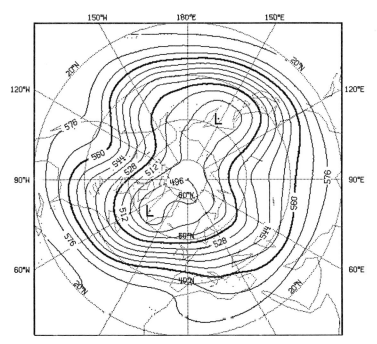

Fig. 2. 500 mb flow January 1981 (top) and ensemble average of all 10-day forecasts for January 1981 (bottom)

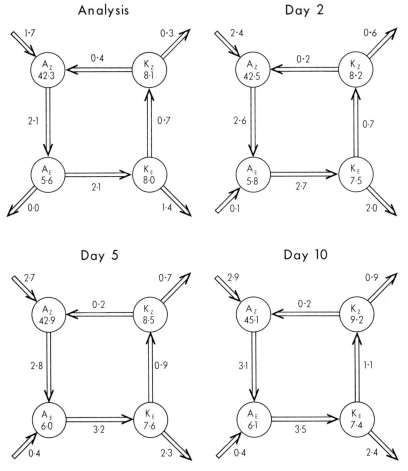

Fig. 3. Global energy diagram for January 1981 as calculated from operational analyses and forecasts. The energies (inside the circles) are given in $10^2 kj/m^2$ while the energy transformations are given in watts/m^2

calculation from the analyses is clearly an underestimation due to the spin-up process, while the values towards the end of the forecast are very likely too large. Probably the figures valid for day 2 (2.6 Watts/m^2) is the most correct. It is interesting to note the figures which describe the generation of eddy available potential energy. When calculating this term from initial data or from short range forecasts, it is found that heating processes destroy eddy available potential energy mainly due to the heating of deep cold air masses over oceans which destroy eddy available potential energy. However, heating due to release of latent heat in the warm air masses increases the eddy available potential energy and hence counteracts this destruction. Because of the fact that the spin-up time for latent heat release is a few days, this contribution to the eddy available potential energy is underestimated and cannot be accurately assessed from the analyses.

Figure 3 illustrates another interesting aspect of the model and its deviation from nature. The intensification of the energy cycle appears to be caused by a too high generation of available potential energy by non-adiabatic processes. A contributing factor to this is possibly related to an incomplete destruction of available potential energy by turbulence and convection in the cold air masses. That this is the case is suggested by a mid troposphere cooling at middle latitudes in the winter. Figure 4 shows the heat balance over a 10-day forecast averaged over the whole globe. The radiative flux is cooling the atmosphere more than is compensated for by the surface fluxes of latent and sensible heat. This gives a cooling of the whole atmosphere by 18.8 Watts/m^2 or 1.6°C in 10 days. The heat balance is negative up to about 50 days, at which time the average global temperature has fallen by about 4°. No further cooling appear to take place hereafter. The gradual cooling and adjustment of the heat balance is shown in Figure 5. The cooling of the atmosphere has a typical vertical profile with a maximum around 500 mb and in the lower troposphere. The reason for this is not yet

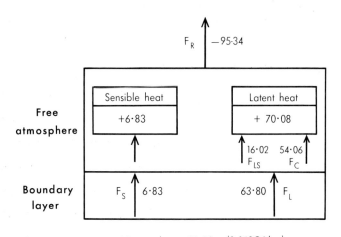

Fig. 4. Global atmospheric heat balance averaged over a 10-day forecast. F_R indicates the total radiation flux, F_S the sensible heat flux and F_L the latent heat flux. F_{LS} and F_C denotes heating by large scale precipitation and by convective precipitation respectively.

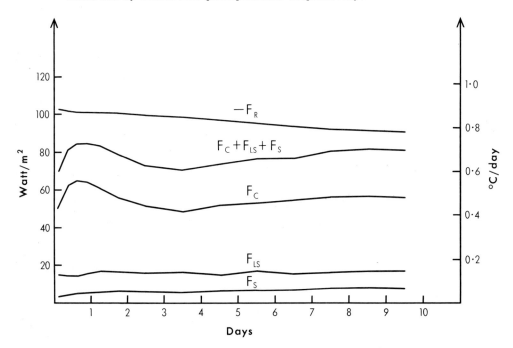

Fig. 5. Global fluxes of F_R (total radiation flux), F_S (sensible heat flux), F_{LS} (heating by large scale precipitation) and F_C (heating by convection) over a 10-day forecast. Note the spin-up effect (in particular for F_C) and the tendency to over-shooting during the first day

fully understood and intense research effort is under way to understand this problem.

6. CONCLUSIONS

We have demonstrated that substantial improvements have taken place in numerical forecasting since the start of numerical weather prediction in 1950. The improvements can be seen as successively better short range forecasts and a gradual extension of the length of useful forecasts up to about 6 to 7 days for the northern hemisphere winter. There are indications that this perhaps can be extended with another couple of days when the systematic deficiencies of present models have been eliminated. However, this may take a considerable time. There is still no realistic alternative to the "brute force approach" and future models are likely to improve through a systematic and meticulous improvement of all aspects of the forecasting system. We can therefore recommend the same general strategy as six years ago.

An alternative approach can possibly be considered for extended range forecasts where the predictability of the individual weather systems has ceased. As has been shown by Miyakoda (pers.inform.) and Shukla (1982) there are indications that the useful predictions of time averages can be extended even further, possibly up to a month. It seems that the most logical approach to such very long integrations would be to apply a simple kind of Monte-Carlo prediction using a sequence of initial states, say over a few days, to produce an ensemble of predictions.

7. ACKNOWLEDGEMENT

The author acknowledges the assistance of H.Savijärvi in undertaking the prediction experiments with the barotropic model.

REFERENCES

Arakawa, A. and V.R.Lamb,1977 Computational design of the basic dynamical processes of the UCLA general circulation model. Methods in Computational Physics, Vol.17 J.Chang, Ed., Academic Press, 337 pp.

Arakawa, A. and W.H.Schubert 1974 Interaction of a cumulus cloud ensemble with the large-scale environment. J.Atmos.Sci., 31, 674-701.

Baer, F. 1977 Adjustment of initial conditions required to suppress gravity oscillations in non-linear flows. Beitr.Phys.Atmos., 50, 350-366.

Bengtsson, L. 1975 4-dimensional assimilation of meteorological observations, GARP Publiction Series No.15, World Meteorological Organizationn, 76pp.

Bengtsson, L. and A.Lange 1982 Results of the WMO/CAS NWP data study and intercomparison project for forecasts for the Northern Hemisphere in 1979-80. CAS Working group on weather prediction research. WMO

Bengtsson, L. 1983 Results of the Global Weather Experiment. Lectures presented at the thirty-fourth session of the WMO Executive Committee - WMO No.610.

Bengtsson, L. and Simmons, A.J., 1983 Medium range weather prediction - Operational experience at ECMWF. Large-scale dynamical processes in the atmosphere. Ed, B.J. Hoskins and R.P.Pearce, Academic Press.

* Burridge, D.M. 1979 Some aspects of large scale numerical modelling of the atmosphere. Proceedings of ECMWF Seminar on Dynamical Meteorology and Numerical Weather Prediction, Vol.2,1-78.

* Burridge, D.M. and J.Haseler 1977 A model for medium range weather forecasting - Adiabatic formulation. ECMWF Tech.Rep.No.4, 46pp.

Eliasen, E., Machenhauer B., and Rasmussen, E., 1970 On a numerical method for interpretation of the hydrodynamical equations with a spectral representation of the horizontal fields. Report No.2., Institut for theoretisk meteorologi, University of Copenhagen.(Available from the Institut for theoretisk meteorologi, University of Copenhagen).

* ECMWF, 1975 Seminars on scientific foundation of medium range weather forecasts, ECMWF 713pp.

* Geleyn, J.-F. 1980 Some diagnostics of the cloud/radiation interaction in ECMWF forecasting model. Workshop on radiation and cloud-radiation interaction in numerical modelling ECMWF, 135-162.

* Girard, C. and Jarraud M., 1982 Short and medium range forecast differences between a spectral and a grid point model. An extensive quasi-operational comparison. ECMWF Technical Report No.32.

Kuo, H-L. 1974 Further studies of the parameterization of the influence of cumulus convection in large-scale flow. J.Atmos.Sci., 31, 1232-1240.

Lettau, H.H. and Davidson, B. 1957 Exploring the Atmosphere's First Mile, Vol.2, Pergamon Press, New York, 578pp.

Lorenc, A.C. 1981 A global three-dimensional multivariate statistical interpolation scheme. Mon.Wea.Rev., 109, 701-721.

* available from ECMWF, Shinfield Park, Reading, Berks, UK.

Lorenz, E. 1982 Atmospheric predictability experiments with a large numerical model, Tellus, 34, 505-513.

Louis, J.-F. 1979 A parametric model for vertical eddy fluxes in the atmosphere. Bound.Lay.Met., 17, 187-202.

Machenhauer, B. 1977 On the dynamics of gravity oscillations in a shallow-watermodel with application to normal mode initialisation. Beitr.Phys.Atmos., 50, 253-271.

Mellor, G.L. and Yamada, T. 1974 A hierarchyof turbulence closure models for planetary boundary layers. J.Atm.Sci., 31, 1791-1806.

Mintz, Y., 1983 The sensitivity of numerically simulated climates to land-surface boundary conditions. To be published in Global Climate, Ed. J.Houghton, Cambridge University Press.

Miyakoda, K., G.D.Hembree, R.F.Strickler and I.Shulman 1972 Cumulative results of extended forecast experiments. I. Model performance for winter cases. Mon.Wea.Rev., 100, 836-855.

* Miyakoda, K. 1975 Weather forecasts and the effects of the sub-grid scale processes. Seminar on scientific foundation of medium range weather forecasts, ECMWF, 380-593.

Miyakoda, K. and Sirutis, J. 1977 Comparative integration of global models with various parameterized processes of sub-grid scale vertical transports: Description of parameterization. Beitr.Phys.Atmos., 50, 445-487.

Rowntree, P.R. and Bolton, J.A., 1978 Experiments with soil moisture anomalies over Europe. The GARP Programme on Numerical Experimentation: Research Activities in Atmospheric and Ocean Modelling. Report No.18. WMO/ICSU Geneva, August 1978, p.63. (To be published in Proceedings of the Symposium on the Global Water Budget Oxford, August 1981).

Orszag, S.A., 1970 Transform method for calculation of vector-coupled sums: Application to the spectral form of the vorticity equation. J.Atmos.Sci., 27, 890-895.

Sadourny, R. 1975 The dynamics of finite difference models of the shallow-water equations. J.Atmos.Sci., 32, 680-689.

Shukla, J. 1982 Predictability of monthly means. Part I: Dynamical predictability. Part II: The influence of the boundary forcing. Proceedings of ECMWF Seminar: Problems and prospects in long and medium range weather forecasting. (This volume).

Shukla, J. and Mintz, Y. 1982 The influence of land-surface evapotranspiration on the earth's climate. Submitted to Science.

Staff Members, University of Stockholm 1954 Results of forecasting with the barotropic model on an electronic computer (BESK) Tellus, 6, 139-149.

Walker, J.M. and Rowntree, P.R., 1977 The effect of soil moisture on circulation and rainfall in a tropical model. Quart.J.R.Met.Soc., 103, pp.29-46.

---

* available from ECMWF, Shinfield Park, Reading, Berks, UK.

Current Problems in Medium Range Forecasting at ECMWF: Model Aspects

A. J. SIMMONS

Abstract

Some characteristic model errors revealed by two years of operational forecasting are described. Systematic mean errors are presented, and a brief discussion of errors in the treatment of transient mid-latitudinal disturbances is also given. The performance of the model for the Tropics is mentioned. Some particular questions relating to numerical techniques, resolution and parameterizations are also discussed.

1. INTRODUCTION

In this contribution we discuss some of the problems in atmospheric modelling of particular current relevance for medium-range weather prediction at ECMWF. The material presented here may be generally divided into two inter-related parts. The first comprises a presentation of some of the systematic deficiencies of the ECMWF forecasting model that have been revealed over the first two years of its operational implementation. The second concerns some of the unanswered questions which have arisen either from the results of research experiments or in the planning of future work.

An outline of the operational forecast model is given in Fig. 1. The model uses a finite-difference scheme based on a staggered grid of variables known as the C-grid (Arakawa and Lamb, 1977). Choice of this grid was made mainly because of its low computational noise and the ease of implementation of a semi-implicit time scheme. Following the work of Arakawa (1966) and Sadourny (1975), the finite difference scheme was designed to conserve potential enstrophy during vorticity advection by the horizontal flow. Further detail has been given by Burridge and Haseler (1977), and Burridge (1979).

CHARACTERISTICS OF THE ECMWF OPERATIONAL GLOBAL GRID POINT MODEL AS AT AUGUST 1981

SIGMA LEVELS

0.025 (σ_s)
0.077
0.132
0.193
0.260
0.334
0.415
0.500
0.589
0.678
0.765
0.845
0.914
0.967
0.996 (σ_{15})

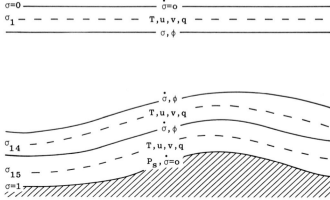

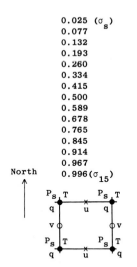

Vertical and horizontal (latitude-longitude) grids and dispositions of variables in the ECMWF grid-point model. Vertical coordinate: $\sigma = p/p_s$

Independent variables	$\lambda, \varphi, \phi, t$
Dependent variables	T, u, v, q, p_s
Grid	Staggered in the horizontal (Arakawa C-grid). Uniform horizontal (regular lat/lon). Resolution: $\Delta\lambda = \Delta\phi = 1.875$ degrees lat/lon. Non-uniform vertical spacing of the 15 levels (see above).
Finite difference scheme	Second order accuracy.
Time-integration	Leapfrog, semi-implicit ($\Delta t = 15$ min) (time filter $\nu = 0.05$)
Horizontal diffusion	Linear, fourth order (diffusion coefficient = 4.10^{15})
Earth surface	Albedo, roughness, soil moisture, snow and ice specified geographically. Albedo, soil moisture and snow time dependent.
Orography	Averaged from high resolution data set.
Vertical boundary conditions	$\dot{\sigma} = 0$ at $p = p_s$ and $p = 0$.
Physical parameterisation	(i) Boundary eddy fluxes dependent on roughness length and local stability (Monin Obukov) (ii) Free-atmosphere turbulent fluxes dependent on mixing length and Richardson number (iii) Kuo convection scheme (iv) Full interaction between radiation and clouds (v) Full hydrological cycle (vi) Computed land temperature, no diurnal cycle (vii) Climatological sea-surface temperature

Fig. 1.

Vertical and horizontal resolutions were selected, within overall computational constraints, to provide a reasonable description of the fundamental large-scale instabilities, some representation of the stratosphere, and an explicit boundary-layer structure. The related parameterization scheme (Tiedtke et al, 1979) describes the interactions thought to be of importance in the medium range, including a full hydrological cycle, a relatively detailed stability-dependent representation of boundary and free-atmospheric turbulent fluxes, and an interaction between the radiation and model-generated clouds.

2. SYSTEMATIC MEAN ERRORS

An important part of the total model error is revealed by averaging forecast errors over a number of cases. These "systematic errors" have been calculated routinely using, for convenience, monthly means. They characteristically grow in amplitude throughout the forecast period, and their general similarity towards the end of this period to errors in the model climatology revealed by integration over extended periods indicates that these errors represent a gradual drift from the climate of the atmosphere towards that of the model. The rate of this drift is found to vary from case to case, but the overall error associated with it appears to be independent of the initial data.

The importance of this climatological component of the forecasts was recognized by Miyakoda et al (1972) in their early series of medium-range forecasts. Two particular errors noted by them were a general cooling of the troposphere, and too low values of the 500 mb height, and to a lesser extent the 1000 mb height, over the North-Eastern Pacific and Atlantic Oceans. Similar, though larger, height errors were found in a trial series of forecasts using February cases carried out at ECMWF by Hollingsworth et al (1980). These authors also estimated that correction of model deficiencies leading to this systematic error might directly lead to an increase of some 20% in objective estimates of the period of useful predictability. Related improvements in transient features might also be expected.

Operational experience has essentially confirmed the findings of Hollingsworth et al. We here restrict attention to the extratropical Northern Hemisphere, and present systematic temperature and height errors towards the end of the forecast period for January 1981. This month was characterized by a persistent, relatively strong circulation, and the systematic error was particularly large.

Meridional cross-sections of errors in the zonal-mean temperature and zonal-mean zonal geostrophic wind, averaged from day 7 to day 10 of the forecasts, are shown in Fig. 2. An overall cooling of the troposphere, by a maximum of 3K at 60°N and 500 mb is evident. The stratosphere was cooled by a substantially larger amount during this month, although this particular error has subsequently been reduced by ensuring an initial pressure-to-sigma interpolation of temperature which is more consistent with the formulation of the model's radiative parameterization. The tropospheric cooling shows little latitudinal variation below 500 mb, and consistent with this the zonal-mean wind error is largely independent of height in this region. The model surface flow is generally stronger than in reality. At upper levels the zonal-mean subtropical jet is displaced poleward, and its strength decreases less rapidly with height than is observed in the stratosphere.

The zonal-mean temperature shows little error at 850 mb, but this disguises a substantially larger error which is revealed by study of the geographical distribution of temperature. Maps of the day 10 temperature error at 500 and 850 mb, and of the 500 and 1000 mb height error, are presented as Fig. 3. Looking first at the height field we see very similar error patterns at 1000 and 500 mb, with distinct centres of low pressure over the North-Eastern

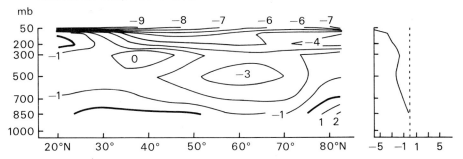

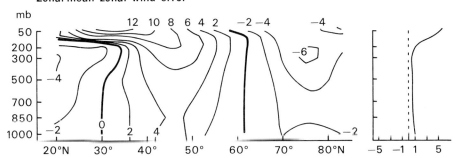

Fig. 2. Meridional cross-sections of zonal-mean temperature error (K) and zonal wind error (ms^{-1}) for the Northern Hemisphere calculated as averages from day 7 to day 10 of all forecasts from January 1981. Vertical profiles of the error averaged from 20°N to 82.5°N are shown in the right-hand plots

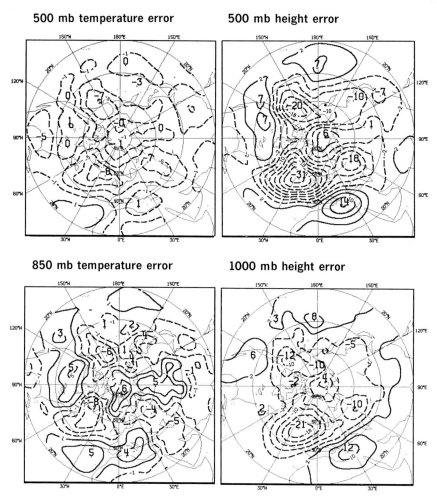

Fig. 3. Monthly-mean error maps for day 10 forecasts of temperature at 500 mb (upper left) and 850 mb (lower left), and height of the 500 mb (upper right) and 1000 mb (lower right) pressure surfaces for January 1981. Contour intervals are 2K and 4dam

Atlantic and Pacific Oceans, and a third centre at 60°E. The amplitude of the error increases with height, and consistent with this, areas of too low temperature tend to coincide with the areas of too low pressure, particularly at 500 mb. Elsewhere the 500 mb temperature error is small, but regions of substantially too warm 850 mb temperature are evident over North America, Siberia and Southern Europe. The temperature error over the latter region is atypical, but warm 850 mb temperatures over the other two areas occur commonly, and become even more pronounced over longer periods of integration. The general distribution of temperature error implies areas of quite erroneous static stability, and a serious impact on transient wave behaviour seems likely. At 850 mb the error may be related to the erroneous horizontal advection implied by the error in the height field.

The error map of the 1000 mb height field shown in Fig. 3 corresponds to erroneously deep Aleutian and Icelandic surface lows. The phase change with height of the atmospheric standing wave pattern is, however, such that the negative centres at 500 mb correspond to an underestimate of the climatological ridges which occur in reality over the North-Eastern Atlantic and Pacific. The model thus exhibits a tendency to predict a too-zonal time-mean flow at upper levels (Fig. 4). Over Europe this is seen in a southward displacement of the jet stream, associated with which is a synoptically-important southward displacement of cyclone tracks.

Examining maps corresponding to that shown in Fig. 3 shows the detailed pattern to vary from month to month, but centres of erroneously low pressure are almost invariably seen over the two ocean regions during the winter months. Maxima are generally less than the 31 dam illustrated for January 1981, but typically reach vaues around 20 dam at 500 mb. Error patterns appear more variable in summer and amplitudes are some 50% lower. 500 mb error maxima greater than 20 dam, and a general weakening of the monthly-mean

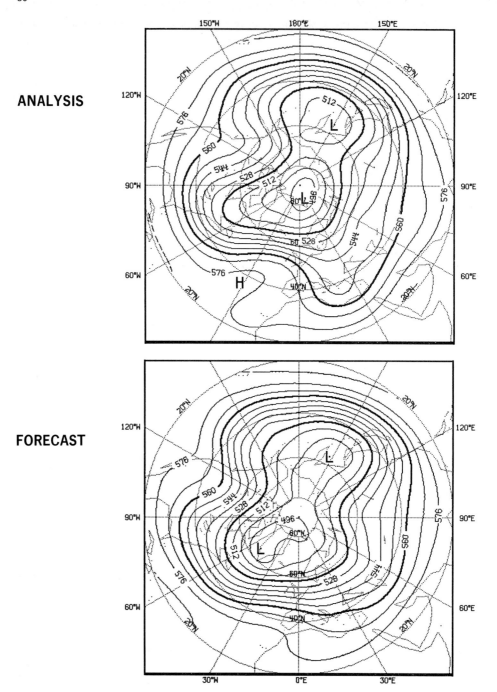

Fig. 4. Monthly-mean 500 mb height maps for the period 11 January to 10 February 1981. The upper map shows the mean of analyses and the lower map the mean of the day 10 forecasts verifying in this period. The contour interval is 8 dam

standing wave pattern, may also be seen in day 10 forecasts for the Southern Hemisphere.

There are other points of interest concerning the systematic height error. In its equivalent-barotropic structure and location of amplitude maxima it tends to resemble actual anomalies of the atmospheric circulation such as discussed by Wallace and Gutzler (1981). In many respects it is by no means unique to the forecast model discussed here (Bengtsson and Lange, 1981, Wallace and Woessner, 1982), and experience at ECMWF confirms that of Manabe et al (1979) who found this error to increase with increasing horizontal resolution in climate simulations. Solution of this particular model problem is thus of importance both for medium-range prediction and for the numerical simulation of climate. Some sensitivity of the error to the parameterization of turbulent fluxes has been found, and further investigations are being actively pursued.*

3. ERRORS IN THE FORECAST OF TRANSIENT MID-LATITUDE DISTURBANCES

A substantial amount of information concerning the treatment of transient mid-latitude disturbances by the ECMWF forecasting system is contained in the synoptic assessments received from Member States. It is not within the scope of this paper to review these, but reference may be made to a forthcoming article in the Centre's Meteorological Bulletin series, No.2.1/1. Here, by way of example, we use anomaly correlations of height averaged over the extratropical Northern Hemisphere for the forecasts of May 1981, to select examples of relatively good and bad forecasts at days 4 and 7. Figure 5 presents maps of the 1000 mb height field for the 4-day forecasts from 12th and 19th May, together with the verifying analyses. Corresponding 7-day forecasts for 500 mb are shown in Fig. 6. The forecast from the 12th is, according to our chosen objective measure, the best of the month at both day

---

*A recent study by Wallace et al (1983) has shown the pattern of short-range forecast error to be suggestive of an underestimation of orographic forcing, and systematic medium-range errors were found to be substantially reduced by use of a higher "envelope" orography in the forecast model

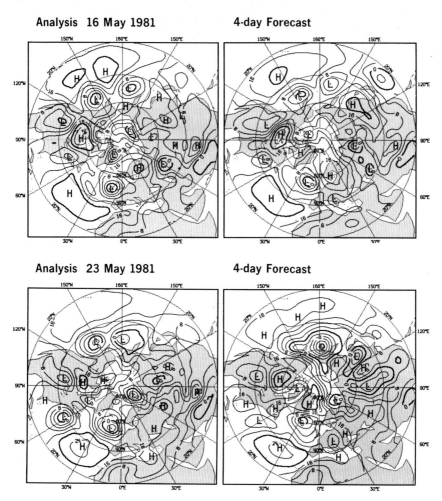

Fig. 5. Maps of 1000 mb height for the extratropical Northern Hemisphere. The left-hand plots are analyses for 16 May, 1981 (upper) and 23 May, 1981 (lower), while the right-hand plots are 4-day forecasts verifying on these two dates. The contour interval is 4 dam

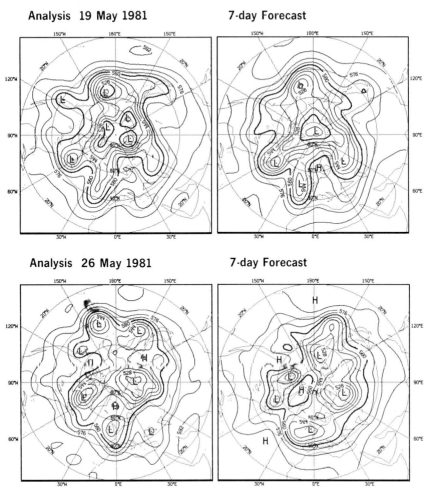

Fig. 6. As Fig. 5, but for 7-day forecasts and the 500 mb height. The contour interval is 8 dam

4 and day 7, whereas that from the 19th is close to the worst of the month at both days. Anomaly correlations averaged over the extratropical Northern Hemisphere are 79% and 62% respectively for the day 4 forecasts at 1000 mb, and 62% and 32% for the two day 7 forecasts at 500 mb.

The surface synoptic situations were very similar on the 16th and 23rd May. Particular common features to note are the two lows in the Central and Eastern Pacific, the low to the west of Ireland and the developing low near the eastern coast of North America, although the latter is of different amplitude on the two days. Despite this similarity, the two day 4 forecasts are quite different. Although the better of the two has errors of detail, it has clearly forecast with reasonable accuracy the positions and intensities of the four lows in question.

The forecast from 19th May, however, exhibits at day 4 a number of errors which may commonly be seen at later stages of the ECMWF forecasts. Particularly worthy of note is the Pacific sector, where the forecast has produced not two distinct lows but what appears closer to one, large-scale and large-amplitude depression. Conversely, the development of the low near 60°W has been substantially underestimated, while the phase of the low to the west of Northern Europe has also been poorly forecast.

Overdevelopment of depressions, such as illustrated near 180°E in the above example, tends to occur most commonly over the Central and Eastern Pacific, and over the Eastern Atlantic and Northern Europe. In the mean this appears as the intensification and eastward shift of the Aleutian and Icelandic lows discussed in the preceding section. In contrast, late or inadequate development of new disturbances over the Western Atlantic occurs in a number of cases, and an underestimation of the phase speed of rapidly-moving lows is frequently observed, although in the forecast from 19th May the low centred near 30°W had an origin significantly different from that of the analyzed low centred further towards the east.

In view of the deficiencies of the 4-day forecast from 19th May it is not surprising that the 7-day forecast exhibits little skill. Figure 6 shows that while the analyzed and forecast charts bear some overall resemblance (corresponding to the anomaly correlation of 32%) there is substantial error at most longitudes. In contrast, all main troughs exhibit a reasonably accurate position and amplitude in the 7-day forecast from a week earlier, although some detail is lost over the Pacific and the west of North America.

4. TROPICAL FORECASTS

Although the forecasts for the Tropics have not been evaluated in as extensive an objective and subjective way as those for the extratropical Northern Hemisphere, there is no doubt that at present their accuracy and usefulness is very substantially less. Indeed, standard deviations of forecast winds show, in the mean, no improvement over persistence in the lower troposphere, while synoptic assessment reveals that some distinct errors in the low-level flow develop quite systematically in the earliest stages of the forecast. Maintenance of the quasi-stationary regional circulations of the tropical atmosphere is evidently a distinct modelling problem, and in the zonal-mean there is a partial suppression of the Hadley circulation and an underestimation of tropical precipitation by some 20%.

Despite these deficiencies, individual cases of quite accurate forecasts of transient behaviour may also be found. For the longer range there are arguments to suggest that the predictability of temporal and spatial means of the tropical atmosphere may be higher than that of middle latitudes (Shukla, in this Volume) but such studies have not as yet been carried out at ECMWF.

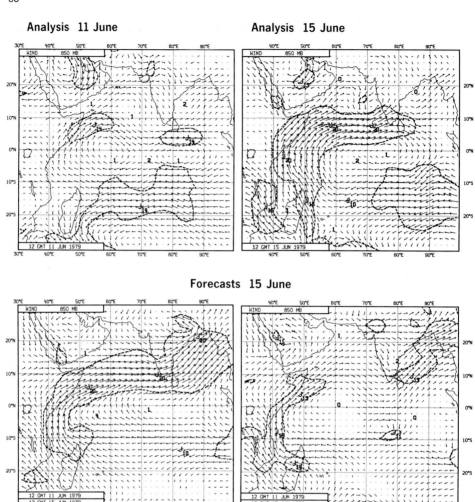

Fig. 7. Analyses of 850 mb wind for 11 June (upper left) and 15 June (upper right), and corresponding 4-day forecasts for 15 June using the Arakawa-Schubert (lower left) and Kuo (lower right) convection schemes. Flow maxima and minima are marked in ms^{-1}

The indication from both subjective and objective assessments of the tropical forecasts is that there are serious deficiencies in the parameterization of convection, and a substantial effort to understand and correct these deficiencies is currently being made. As an example of the sensitivity that can be found, we show in Fig. 7 two 4-day forecasts of the 850 mb wind over the Indian Ocean and bordering areas. These forecasts (which used FGGE rather than operational data) cover a period starting from 11 June 1979 which was marked by the onset, rather later than normal, of the south- west monsoon. The two differ only in their parameterization of convection, one using the scheme of Kuo (1974) adopted for operational forecasting and the other the Arakawa-Schubert (1974) scheme. The latter evidently produces a quite different forecast, and in fact a very much better representation of the development of the strong monsoon flow over the Arabian Sea. Just one experiment cannot of course be used to draw firm conclusions as to which of the convection schemes is the better (and indeed, the extratropical forecasts in this case were slightly the better using the Kuo scheme), but the sensitivity of the forecast is worth noting. Other studies have in addition shown sensitivity to the prescription of soil moisture and orography, results in general agreement with those found elsewhere (Rowntree, 1978). Overall, it seems that the tropical forecasts respond more quickly and acutely to defects in the model than do forecasts at middle and high latitudes.

5. COMPARISONS OF GRID-POINT AND SPECTRAL TECHNIQUES FOR THE HORIZONTAL REPRESENTATION

In designing a new forecasting system, a choice has to be made concerning the discretization techniques to be adopted in normal use of the system. Research on this topic at ECMWF has initially concentrated on a comparison of finite-difference and spectral methods for the horizontal representation. Detailed results of an extended experiment comparing forecasts performed once per week for a complete year have been given by Girard and Jarraud (1982). In this experiment, the operational grid-point model forecasts (which used a

1.875°, or "N48" resolution) were compared with spectral forecasts using triangular truncation at total wavenumber 63 (T63), these two models requiring a similar amount of computing resources.

Although the models often gave a very similar forecast, some clear differences in overall performance were found. An indication of this is given by Fig. 8, while Fig. 9 presents one example (out of by no means few) of a markedly better local forecast by the spectral model. The occurrence of such differences in the medium range raises the question as to the extent of any future improvement that might be obtained by refinements in numerical technique or resolution.

Although many of the systematic errors noted in previous sections occur for the T63 spectral model as well as for the operational finite-difference model, some reduction in a number of these errors has been found. One example is the overall cooling, which has been found to be less in the series of spectral forecasts. A second, for which statistics are presented in Table 1, is the tendency to underestimate the phase-speed of rapidly-moving lows. Table 1 shows phase speeds to be generally better represented by the spectral model, at least in the short-range (for which an unambiguous identification of analyzed and forecast lows was possible).

6. HIGHER RESOLUTION

A question which naturally arises during the course of work at a numerical forecasting centre is that of the extent to which forecasts might be improved by increases in horizontal or vertical resolution. One approach to answering this question is to extrapolate using results from the current resolution and those obtained using lower resolution. In this respect, experiments with the ECMWF spectral model show a clear improvement to result from increasing horizontal resolution from T40 to T63 (Jarraud, Girard and Cubasch, 1981). A

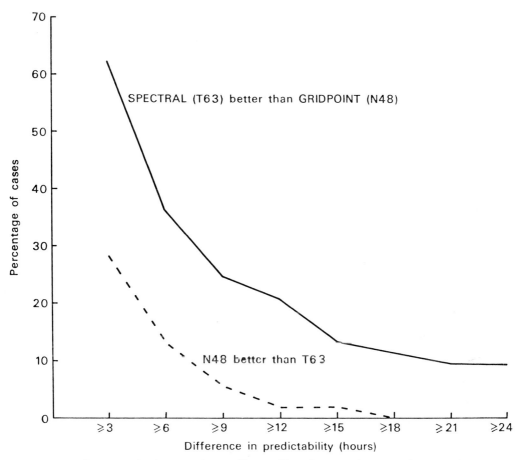

Fig. 8. The difference in predictability (measured by the length of the forecast period for which the anomaly correlation of the 1000 mb height over the extratropical Northern Hemisphere remains above 60%) between spectral (T63) and grid-point (N48) models. Results are expressed in terms of the percentage of cases for which one or other model gave better results

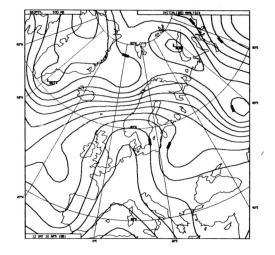
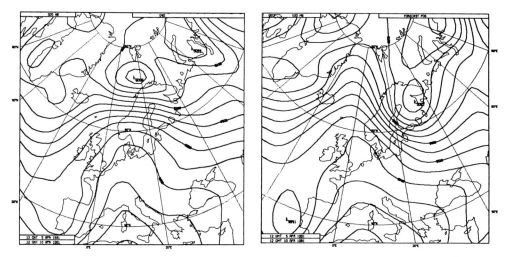

Fig. 9. The analyzed 500 mb height for 10 April, 1981 (upper) and 5-day forecasts for this date by the T63 spectral model (lower left) and the N48 grid-point model (lower right)

Table 1. Errors in the displacement (in degrees longitude) of surface lows between day 1 and day 2 of the forecasts for spectral (T63) and grid-point (N48) model forecasts.

Displacement (D)	Cases	Error (Degrees) T63	N48
$D < 5°$	64	+.6	+1.0
$5° \leq D < 10°$	39	+.3	+.2
$10° \leq D$	89	-1.8	-2.6
$15° \leq D$	44	-1.8	-3.3
$20° \leq D$	16	-2.9	-4.5

more convincing approach is actually to carry out higher resolution forecasts, although computational limitations often inhibit a comprehensive experimental programme.

Only very limited evidence of the influence of higher resolution is currrently available for the ECMWF forecasting models, and it would be unwise to attempt to draw any firm conclusions at this stage. Figure 10 illustrates how the intensity of a rapid, initially small-scale, development is captured better with increasing horizontal resolution, while Fig. 11 shows, from the same forecast, a much-improved indication of the flow over the North Eastern Atlantic from the highest resolution (T96) forecast. It should be noted that all forecasts shown in Figs. 10 and 11 were performed from an analysis produced using the N48 grid-point model in the data assimilation. The sensitivity of forecasts of relatively small-scale developments such as shown in Fig. 11 is emphasized by Fig. 12, which compares two (grid-point model) forecasts from analyses which differed not in the content of the data sets used, but rather by the refinements in analysis technique made over the course of a year or so's development of the ECMWF system. It appears that in view of results such as these, due attention should be paid to the data assimilation in the planning of higher resolution experiments. In particular, use of the higher resolution model in the data assimilation cycles is probably necessary for a reliable indication of any benefits of higher resolution.

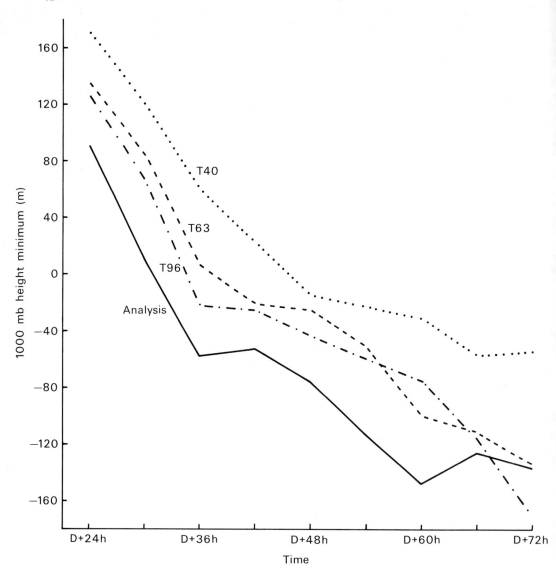

Fig. 10. Forecasts (from 12GMT, 18 Feb 1979) of the rapid development of the "President's Day Storm" using three different spectral resolutions

7. PARAMETERIZATION

In our preceding discussion of model deficiencies a number of references has been made to aspects of the parameterization. Thus we have noted some sensitivity of the systematic height error to the parameterization of turbulent fluxes, sensitivity of the tropical forecasts to the parameterization of convection, and sensitivity of the stratospheric temperature error to the parameterization of radiation. A detailed

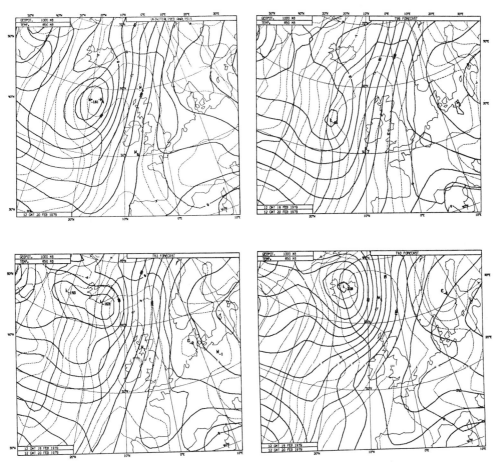

Fig. 11. Maps of 1000 mb height (contour interval 4 dam) and 850 mb temperature (contour interval 2 K) for 12GMT, 20 February 1079.
Upper left: analysis. Upper right: T96 forecast.
Lower left: T63. Lower right: T40

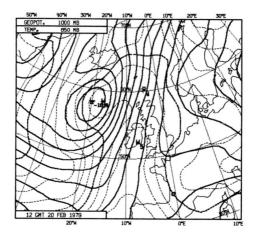
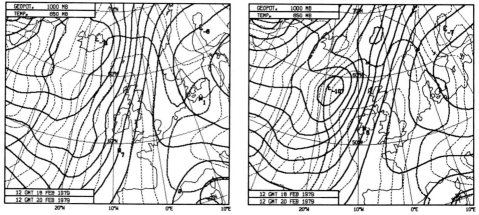

Fig. 12. As Fig. 11, but for the verifying analysis (upper), and for 2-day forecasts from two different initial analyses

presentation of these and other such results is beyond the scope of this contribution, but we mention very briefly below some work on parameterization currently being carried out at ECMWF, and discuss also problems which have recently arisen relating to the interface between the parameterization and the dynamical model.

Concerning first the parameterization itself, effort is being made to evaluate the performance of a variety of representations of convection. These include the previously mentioned schemes proposed by Kuo (1974) and Arakawa and Schubert (1974), a scheme based on some ideas of Lindzen (1981), a scheme developed at ECMWF by Miller and Moncrieff, and the simpler moist convective adjustment (Manabe et al 1965). The major effort in the parameterization of radiation is directed towards inclusion of a diurnal cycle in the model. Preliminary results indicate a quite small impact on forecasts, and further experimentation is continuing. A project to evaluate the use of a higher-order closure scheme in the parameterization of turbulent fluxes has begun following the successful first results obtained by Miyakoda at GFDL.

More generally, evidence appears to point towards a need to pay greater attention to the interaction between the parameterization and the resolved motion of the forecast model. Traditionally, parameterization schemes take input data at discrete levels for a particular atmospheric column, or grid-point, and produce tendencies of the prognostic variables at these discrete levels for the column in question. Such an approach may, however, yield problems.

One example may be found in the parameterization of radiative cooling. For certain distributions of temperature and humidity, increasing the vertical resolution is found to give rise to increasingly large cooling rates at just

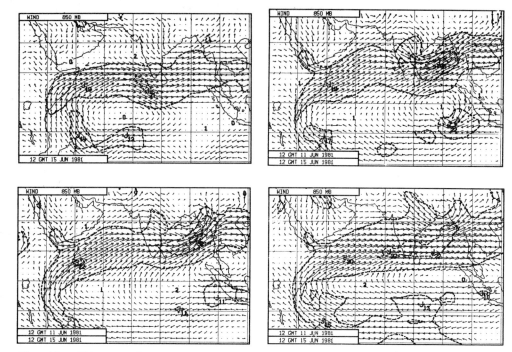

Fig. 13. 850 mb wind analysis for 15 June, 1981 (upper left) and day 4 forecasts for this date using moist convective adjustment (upper right), the operational Kuo convection scheme (lower left), and the Arakawa-Schubert convection scheme (lower right)

one model level. These are not by themselves physically unreasonable, representing as they do a substantial cooling of cloud tops, but the response of the (vertical finite-difference) model to such a distribution of cooling is far from clear.

A second example concerns the current parameterizations of convection. In these schemes, an unstable atmosphere leads to parameterized heating profiles being introduced into the model, but with no associated (dynamical-balancing) change to the wind field. This can result in the convective heating being balanced by adiabatic cooling due to excessive vertical motion, rather than by a stabilizing temperature change. Thus regions of convections may persist unrealistically, with unrealistic related circulation patterns. An example is shown in Fig. 13 in which a too-strong monsoon flow is found (apparently driven by excess convection in the vicinity of Burma) over the Bay of Bengal for three different parameterizations of convection.

8. CONCLUSIONS

In this paper we have presented some model-related aspects of current problems in medium-range forecasting at ECMWF. It must be stressed that the discussion has been by no means comprehensive. We have not, for example, touched upon the problems related to the representation of orography or to related problems of the prediction of precipitation. A number of distinct modelling problems have, however, been illustrated. It is difficult to estimate the exact impact that solution of these problems would have on the accuracy of the Centre's forecasts, but there nevertheless appears to be substantial scope for extending the period for which these forecasts are of practical use.

REFERENCES

Arakawa, A. 1966 Computational design for long-term numerical integration of the equations of fluid motion: two dimensional incompressible flow. Part 1. J.Comp.Phys., 1, 119-143.

Arakawa, A. and V.R.Lamb 1977 Computational design of the basic dynamical processes of the UCLA general circulation model. Methods in Computational Physics, Vol.17, J.Chang, Ed., Academic Press, 337pp.

Arakawa, A. and W.H.Schubert 1974 Interaction of a cumulus cloud ensemble with the large-scale environment. J.Atmos.Sci., 31, 674-701.

Bengtsson, L. and A.Lange 1982 Results of the WMO/CAS NWP data study and intercomparison project for forecasts for the Northern Hemisphere in 1979-80. To be published by WMO.

Burridge, D.M. 1979 Some aspects of large scale numerical modelling of the atmosphere. Proceedings of ECMWF Seminar on Dynamical Meteorology and Numerical Weather Prediction, Vol.2, 1-78.

Burridge, D.M. and J.Haseler 1977 A model for medium range weather forecasting _ Adiabatic formulation. ECMWF Technical Report No.4, 46pp.

Girard, C. and M.Jarraud 1982 Short and medium range forecast differences between a spectral and a grid-point model. An extensive quasi-operational comparison. To be published as ECMWF Technical Report.

Hollingsworth, A., K.Arpe, M.Tiedtke, M.Capaldo and H.Savijarvi 1980 The performance of a medium-range forecast model in winter - Impact of physical parameterizations. Mon.Wea.Rev., 108, 1736-1773.

Jarraud, M., C.Girard and U.Cubasch 1981 Comparison of medium range forecasts made with models using spectral or finite difference techniques in the horizontal. ECMWF Technical Report No.23, 96pp.

Kuo, H-L. 1974 Further studies of the parameterization of the influence of cumulus convection in large-scale flow. J.Atmos.Sci., 31, 1232-1240.

Lindzen, R.S. 1981 Some remarks on cumulus parameterization. Report on NASA-GISS Workshop: Clouds in climate - Modelling and Satellite Observational studies. pp.42-51.

Manabe, S., J.Smagorinsky and R.F.Strickler 1965 Simulated climatology of a general circulation model with a hydrological cycle. Mon.Wea.Rev., 93, 769-798.

Manabe, S., D.G.Hahn and J.L.Holloway 1979 Climate simulation with GFDL spectral models of the atmosphere: Effect of spectral truncation, GARP Publication Series No.22, 41-94.

Miyakoda, K., G.D.Hembree, R.F.Strickler and I.Shulman 1972 Cumulative results of extended forecast experiments. I. Model performance for winter cases. Mon.Wea.Rev., 100, 836-855.

Rowntree, P.R. 1978 Numerical prediction and simulation of the tropical atmosphere. In Meteorology over the Tropical Oceans, Roy.Met.Soc., 278pp.

Sadourny, R. 1975 The dynamics of finite difference models of the shallow-water equations. J.Atmos.Sci., 32, 680-689.

Tiedtke, M., J-F.Geleyn, A.Hollingsworth and J-F.Louis 1979 ECMWF model-parameterization of sub-grid scale processes. ECMWF Technical Report No.10, 46pp.

Wallace, J.M. and D.S.Gutzler 1981 Teleconnections in the Geopotential Height Field during the Northern Hemisphere Winter. Mon.Wea.Rev., 109, 784-812.

Wallace, J.M. and J.K.Woessner 1982 An analysis of forecast error in the NMC hemispheric primitive equation model. To be published in Mon.Wea.Rev.

Wallace, J.M., S.Tibaldi and A.J. Simmons 1983 Reduction of systematic forecast errors in the ECMWF model through the introduction of an envelope orography. Quart.J.Roy.Met.Soc., in press.

Current Problems in Medium Range Forecasting at ECMWF Data Assimilation Scheme

G. J. Cats*

Abstract

A number of examples show that the analysis, made with the ECMWF data assimilation scheme, and/or the subsequent forecast with the ECMWF forecast model are sensitive to many parameters in the assimilation scheme. Because several of those parameters are not very well known (both qualitatively and quantitatively) this sensitivity poses a severe problem to medium range weather forecasting at ECMWF. The sensitivity is particularly related to insufficiency of accurate observational data.

1. INTRODUCTION

A medium range weather forecast is made by running an atmospheric model starting from some initial state of the atmosphere. The initial state is partly described by observations, but the density and accuracy of the observations are not sufficient to define the initial state uniquely. Therefore an analysis is made, which at ECMWF is an update of the atmospheric fields as forecast by the ECMWF model, using the available data. In the analysis scheme many parameters occur, and most of them are not known accurately. Their values have been chosen only to produce reasonable analyses over a large ensemble of cases. A significant problem in medium range weather forecasting, however, is that the forecast is often very sensitive to the precise choice of those parameters.

In this paper a number of examples of such sensitivity in the ECMWF model will be given. The examples relate to two types of sensitivity. The first type shows that the initial state is often sensitive to the parameters used to construct it. It is no surprise that in those cases the forecast is

*Present affiliation: The Royal Netherlands Meteorological Institute, De Bilt, The Netherlands

sensitive as well. Of the second type are the examples where the initial state is hardly affected by the choice of the parameters (at least, in its height, wind, temperature etc. fields) but where the forecast is yet very sensitive to the parameters.

The first type of sensitivity is usually due to the data checking algorithm, which determines whether a datum is correct (in which case it is accepted) or incorrect (in which case it is rejected). A slight change of a parameter in the analysis scheme may cause the rejection of a datum that was previously accepted or vice versa. The examples relating to this mechanism are collected in the next section.

Sensitivities of the first type are sometimes due to other mechanisms, e.g., a small change in static stability in an analysis may trigger convection during a six hour forecast. The (6-hour) forecast fields are used in the next analysis, and if there are no relevant observations in that analysis the analysed fields may alter considerably. Some examples of this kind of sensitivity are shown in the third section.

The second type of sensitivity gives an impression of the accuracy of the analysed fields which is required by the ECMWF forecast model. Examples are described in the fourth section, where we show that the rapid growth and downstream propagations of forecast differences bear a strong similarity to theoretical calculations by Simmons and Hoskins (1979).

For most of the examples a reasonable understanding of the ECMWF data assimilation scheme is useful. Some explanations will be given with each example. A few general remarks on the scheme follow.

The ECMWF data assimilation scheme consist of a repeating sequence of analysis, initialization and six hour forecasts, together called a data assimilation cycle. The forecast, made with the ECMWF forecast model, produces the "first-guess" fields for the next analysis. The analysis mixes information from the first guess field with that from the observations. The observations that are used are from as many sources as practicable. Height, wind, thickness/temperature and humidity observations within a six hour time window centered around the analysis time are considered. The height and wind analysis is three-dimensional, multivariate and based on the optimum interpolation technique (Lorenc 1981). The humidity is analysed with a correction method. Surface parameters (e.g. soil moisture and temperature at the soil surface and at a depth of 1 m) are not analysed yet; instead we use climatological values (apart from soil surface temperature which is taken from the first-guess). The height, wind and humidity are analysed on 15 standard pressure levels, and a vertical interpolation is required to get the analyses on the 15 model σ-levels. The analysed height fields are transformed to temperature fields as required by the model with the hydrostatic equation. In the present operational implementation these fields on σ levels are not changed by the analysis if there are no observations. The initialization scheme has been described by Williamson and Temperton (1981). A summary of the main features of the assimilation scheme is given in Appendix A.

The sensitivity of the forecast to the initial state, and of the initial state to parameters in the data assimilation scheme, renders it impossible to judge the quality of a change in the assimilation scheme from inspection of the meteorological charts of the analysis or from the quality of a small number of test forecasts. Instead, one would require a large sample to assess the average effect of a change. This, however, would greatly exceed available computer and manpower resources. At ECMWF, the quality of the operational forecasts is monitored with objective measures. During the second year of operational forecasting (1980/1981) that quality has improved consistently over that during the first year. Although this improvement might partly be due to the difference in synoptic situations and in data availability, it is felt that it is also at least partly attributable to the large number of changes that have been introduced in the analysis scheme in the beginning of the second operational year. Only now, after a long period of operational quality assessments, we feel justified to say that the analysis scheme changes constituted improvements.

2. SENSITIVITY OF THE ANALYSIS THROUGH DATA CHECKING

2.1 Description of data checking algorithm

In the analysis scheme every observed datum undergoes several checks before it is used for the final analysis. In the mass and wind analysis first checks on internal consistency, agreement with nearby data and agreement with a first guess value are carried out. These are followed by the main check, which is as follows.

The data are grouped into "boxes" of about 660 x 660 km^2. Each datum within a box is compared to a preliminary analysis which avoids the use of any datum already rejected and which also avoids the use of the datum that is being checked. The deviation of the observed value from the preliminary analysis value is expressed in terms of the dimensionless ratio q:

$$q = \frac{(\alpha - a_p)^2}{e_{a_p}^2 + e_\alpha^2} \tag{1}$$

where α is the observed value, a_p the preliminary analysis, e_{a_p} an estimate of the error in that analysis, and e_α an estimate of the observation error.

If for any datum within the box q exceeds q_{lim} (at present 16) the datum with the highest q-value is rejected. This entire process is repeated until for each datum $q \leqslant q_{lim}$. After this has been done for all boxes, the mass and wind fields are analysed using only accepted (non rejected) data.

Sensitivity to any parameter in the analysis scheme reveals itself through the rejection or acceptance of data with the ratio q very close to q_{lim}. If, for example, a small (order ε) change in a parameter, with an associated change in a_p of order ε, reduces q from $q_{lim} + \varepsilon$ to q_{lim}, a datum, previously rejected, will become accepted. Since such a datum is on the border of being rejected, q is large ($q \approx q_{lim}$) and therefore it deviates much from the preliminary analysis. Its inclusion will therefore change the analysis considerably.

The change in the final analysis is estimated as $d = w (\alpha - a_p)$ where w is the weight of the "doubtful" datum, if accepted. Because $q \approx q_{lim}$, $|d|$ is approximately

$$|d| \approx |w| \sqrt{q_{lim} (e_{a_p}^2 + e_\alpha^2)} \qquad (2)$$

The weight w is large if e_α is small, so that in data sparse areas (large e_{a_p}) $|d|$ may become very large. In the next part of this section two examples will highlight this effect in data sparse areas, and one will indicate that in data dense areas there is no such sensitivity. For the first example the actual change that made the rejection scheme treat the data differently is described. The other two cases arose from analyses where many parameters had been changed in their least significant bits, so that these cases indicate the importance of those bits in the computations.

2.2 Examples

2.2.1 First example: choice between two data

The first example, shown in Fig. 1 is from a case where a SHIP and a TEMP report at almost the same position and time have contradictory data for the height of the 1000 mbar level: The TEMP, call sign EREBO, reported -17 m, but the SHIP, call sign EREBO (!), reported 992 mbar for mean sea level pressure

38 SYNOP/SHIP   14 TEMP   12 GMT 7 September 1980

1000 mb First-Guess Geopotential Height

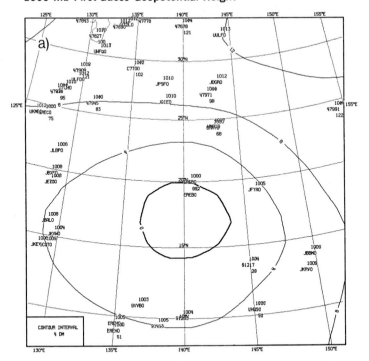

Fig. 1. First guess (a) and two analyses (b,c) for 12 GMT 7 September 1980. In analysis (b) the TEMP report EREBO of -17m for the 1000mb height has been accepted, in analysis (c) the SYNOP report EREBO of 992mb for the mean sea level pressure

12 GMT 7 September 1980
1000 mb Analysed Geopotential Height

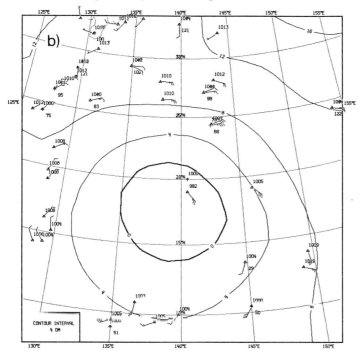

1000 mb Analysed Geopotential Height

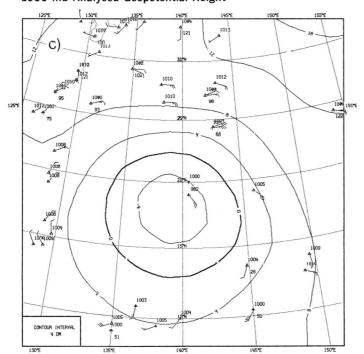

Fig. 1. (cont.)

(MSLP). In the analysis scheme, a correlation between the height observation errors within a single TEMP report is assumed. At present, not enough statistics exist on the root-mean-square height observation errors, let alone on the correlations between these errors. In an analysis using the assumed value of 55% for the correlation between 850 mbar and 1000 mbar height observation errors, the TEMP report is accepted, and the SHIP rejected (Fig. 1b). If that correlation is reduced by 20%, the roles of the TEMP and SHIP reports are swapped (Fig. 1c), resulting in an MSLP change of 4 1/2 mbar.

A case like this is neither exceptional in frequency of occurrence, nor in data distribution, as can be seen already from many other similar contradictory but coincident TEMP and SHIP's in Fig. 1. Experience at ECMWF shows that for whatever change is made in the analysis scheme there is often a point on the globe where the resulting final analysis changes disproportionately.

2.2.2 Second example: data dense area

The second example shows that in a data dense area one can hardly see the effect of the acceptance or rejection of a single datum. In the analysis shown in Fig. 2a, a SYNOP report (longitude 67/1°W, latitude 68.1°S) of MSLP of 978 mbar has been used, whereas it was rejected for the analysis of Fig. 2b. The two MSLP analyses are very similar. It should be borne in mind, however, that other quantities, such as static stability or advection, being first or second derivatives of the height field, may differ enough to yield substantially different 10-day forecasts.

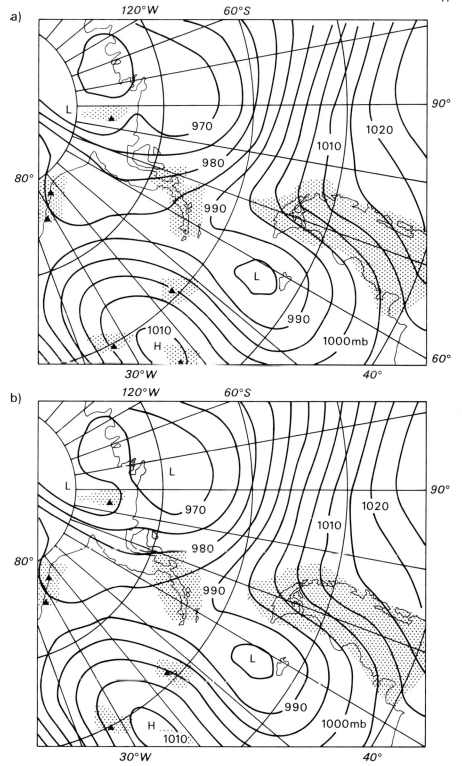

Fig. 2. Analysed mean sea level pressure near the Antarctic Peninsula for 1200 GMT 7 September 1980. In analysis (a) a SYNOP report of 978 mb mean sea level pressure on the peninsula has been accepted, which in analysis (b) has been rejected. Data dense areas have been shaded

2.2.3 Third example: one isolated datum

The third example is again a situation on the southern hemisphere. The MSLP of SHIP GWAN has been rejected by the analysis of Fig. 3a. (This analysis is an operational ECMWF analysis). Because the area has few other data, accepting that datum (Fig. 3b) produces a markedly different analysis of the ridge along 100°W. From both analyses a forecast has been made. The two-day and four-day forecasts are shown in Fig. 4 and 5, respectively, together with the verifying analyses. The stronger ridge of the analysis of Fig. 3b can be followed through the forecast, to produce a strong high over South America. The verifying analyses show it as well, although they have been obtained with the ECMWF operational scheme, that is, after a sequence of assimilation cycles which started from the operational analysis shown in Fig. 3a.

This case is particularly suited to illustrate the effect of one datum on the forecast because the datum is in a data sparse region (so the analysis changes considerably) but the effects in the forecast occur over a data dense area (so that the verifying analyses are reliable).

The two forecasts in this case were made from analyses obtained with slightly different data assimilation schemes. Therefore the analyses differed not only in the treatment of this one datum. In Section 4 an experiment will be described in which the forecasts were run from analyses that differed in that respect only.

3. SENSITIVITY TO ANALYSIS ASPECTS OTHER THAN DATA CHECKING

3.1 Introduction

In this section several steps in the Data Assimilation Scheme will be looked at, in connection with sensitivity of the 10-day forecast to parameters in those steps. The first step in the ECMWF data assimilation scheme is the construction of a first guess field from a six hour forecast from the previous analysis. Because the model keeps temperature on sigma levels (see

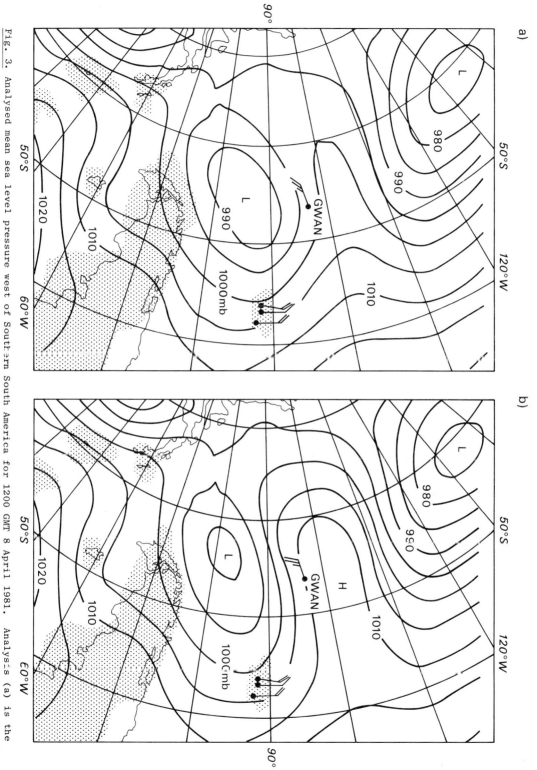

Fig. 3. Analysed mean sea level pressure west of Southern South America for 1200 GMT 8 April 1981. Analysis (a) is the operational analysis; it rejected the mean sea level report of ship GWAN (1011 mb). Analysis (b) is a test, which

12 GMT 10 April 1981 Operational Analysis

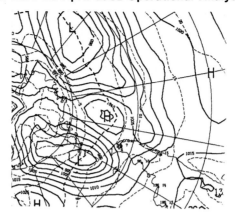

48 h Forecast from 8 April 1981

Operational   Test

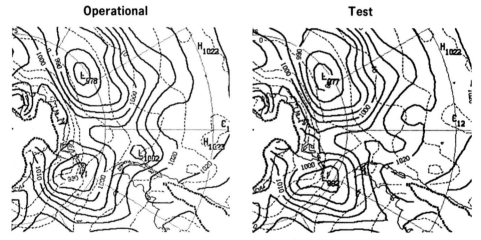

Fig. 4. The 48-hour forecasts from the analyses of Fig. 3 with the verifying analysis (ignore dashed lines)

12 GMT 12 April 1981 Operational Analysis

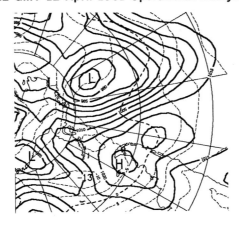

96 h Forecast from 8 April 1981

Operational   **Test**

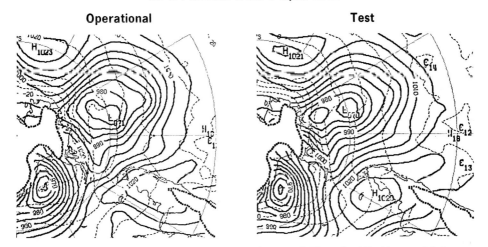

Fig. 5. The 96-hour forecast from the analyses of Fig. 3 with the verifying analysis (ignore dashed lines)

Appendix A), and the analysis calculates height on pressure levels, a vertical interpolation, together with an integration of the hydrostatic equation is required. After the analysis, the reverse interpolation to sigma levels is done. There is a large difference in resolution in the stratosphere between the two coordinate systems (1 sigma and 3 pressure levels above 50 mbar). The interpolation formulae to be used in the stratosphere are not defined by physical considerations, but changing from one formula to another produces widely different analyses. An example of this will be given below.

Apart from mass and wind fields, the model needs an initial prescription of some other data as well. They are also provided by the analysis scheme. An example to indicate the importance of an accurate analysis of one of these parameters is also provided, namely of soil moisture.

After the analysis, the mass and wind fields are initialized. The last step in a data assimilation cycle is a six-hour forecast to produce the first guess fields for the next analysis. These steps will not be treated here, apart from the remark that a small change in each step may lead to a small change in the first guess, therefore to different data rejection and large changes in the analysis.

3.2 Vertical interpolation

Recently both the sigma to pressure interpolation and the pressure to sigma interpolation have been changed in the ECMWF operational forecasting. The combined effect of these two changes is illustrated in Fig. 6.

The 1 day forecast, as shown in Fig. 7a has been made from an analysis in which the sigma to pressure interpolation of geopotential height was carried out assuming the height to have a functional form of a cubic spline in ln p.

T$_{\sigma = 0.025}$ **Operational Analysis**

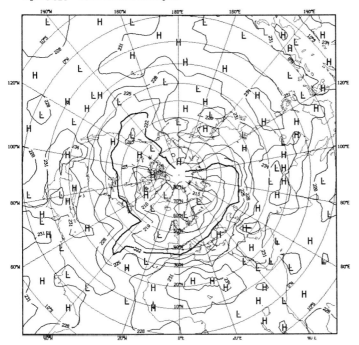

T$_{\sigma = 0.025}$ **Test Analysis**

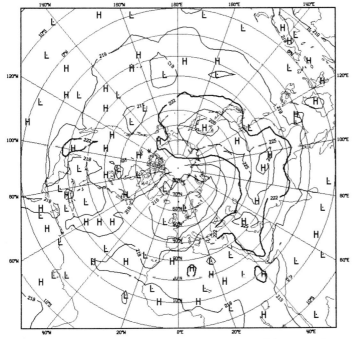

Fig. 6. Top σ-level temperature analysis (K) over the northern hemisphere for 8 April 1981, 12 GMT. Top: operational analysis. Bottom: different pressure to sigma coordinates interpolation

$T_\sigma = 0.025$ **Operational 24 h Forecast**

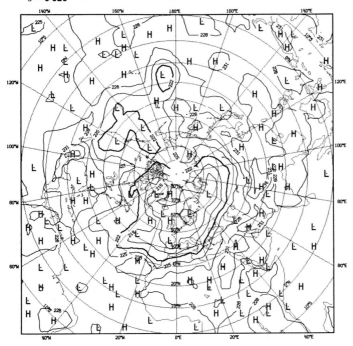

$T_\sigma = 0.025$ **Test 24 h Forecast**

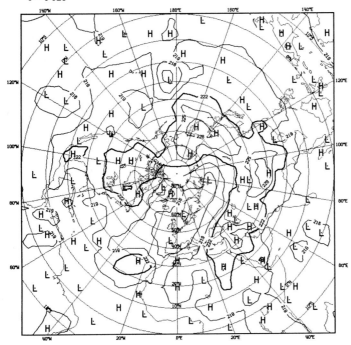

Fig. 7. 24-hour forecasts from the analyses of Fig. 6 (units: K)

The reverse interpolation used a volume weighted mean temperature, in accordance with the hydrostatic equation. The forecast as shown in Fig. 7b, however, followed a data assimilation run of 5 cycles of analyses in which the sigma to pressure interpolation was done with an isothermal atmosphere above $\sigma = 0.050$, and the reverse interpolation used a mass weighted temperature mean, which is more consistent with the radiation scheme in use at ECMWF. Both changes reduce the stratospheric tropical temperatures by a few kelvins. The combined effect after 5 assimilation cycles is up to 15K, and the 30 mbar heights after a 24 hour forecast differ by 200 m in places. Also, the noise level in Figs.6b and 7b is much less than that in Figs.6a and 7a.

This example indicates a sensitivity of the analysis and forecast to some parameters in the data assimilation scheme. There are no indications that those parameters exert influence in the troposphere within 10 days, therefore medium range weather forecasting is not necessarily sensitive to them. The situation in the troposphere is quite different. The vertical interpolation scheme defines static stability and boundary layer structure. It is suspected that the vertical interpolation is a primary source for a sensitivity noted in the ECMWF model, (see Section 4.2). Furthermore, tropical rainfall forecasts have been found sensitive to the formulation of the vertical interpolation of humidity.

3.3 Surface soil moisture specification

The present ECMWF surface soil moisture analysis consists of replacing soil moisture by its climatology after every analysis. A problem, of course, is, that even that climatology is not reliably known yet. Up to now there are no indications that the extratropical forecast is sensitive to the precise specification of the initial surface soil moisture. But, as shown in Fig. 8, over the Indian Ocean forecast 850 mbar winds changed considerably after a

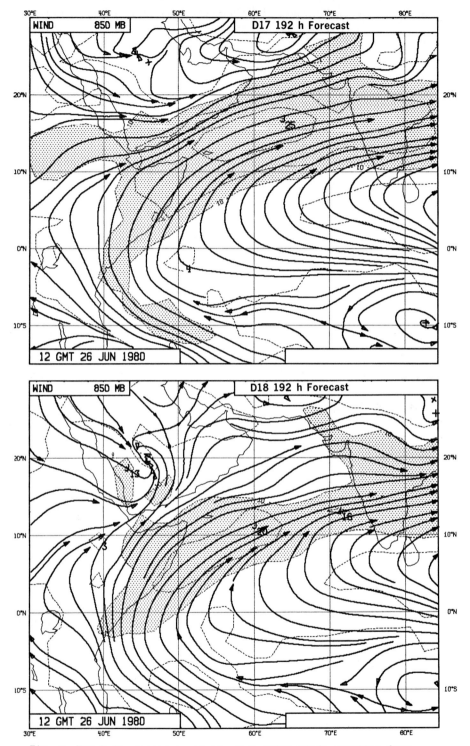

Fig. 8. Two 8-day forecasts of tropical 850mb winds starting from analyses with different soil moisture specification. Numbers in m/s. Shaded are areas with wind speeds in excess of 10 m/s

change in soil moisture specification. The forecast shown in Fig. 9a started from a surface soil moisture analysis that was erroneously wet over Southern Arabia. Neither the northward shift of the low-level jet south of Arabia nor the strong inflow into that jet from North Africa in that "wet" forecast were verified by observations.

4. SENSITIVITY OF THE FORECAST TO THE ANALYSIS

4.1 Introduction

In this section two examples will be presented to indicate that in some synoptic situations the forecast is extremely sensitive to the analysis. The first case concerns the prediction of the "President's-day storm" (February 1979). The second example shows that similar sensitivity may also occur in the tropics; the case presented is the four day forecast of the development of hurricane Orchid (September 1980).

In the second section of this paper two cases demonstrated that the analysis is sometimes sensitive to the data checking. For both cases two analyses were carried out, that only differed in the rejection/acceptance of the data that the second section focused upon. The analyses were followed by 10-day forecasts, and the development of the initial difference through the forecasts will be treated in the fourth subsection. The rapid growth and downstream propagation of the disturbances bear strong similarities to the results of certain theoretical calculations by Simmons and Hoskins (1979).

4.2 The President's day storm.

For one of the observation systems experiments performed by the FGGE-group at ECMWF (N.Gustafsson and J.Pailleux, 1981) two 4-day forecasts were made from 17 February 1979, 1200 GMT. One of the forecasts (label F37, Figs. 9,10 and 11) followed an analysis in a data assimilation experiment in which no satellite wind data were used, the other forecast (label F36) one that used

88

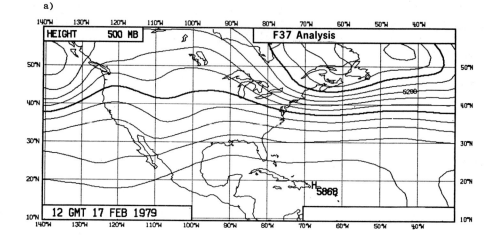

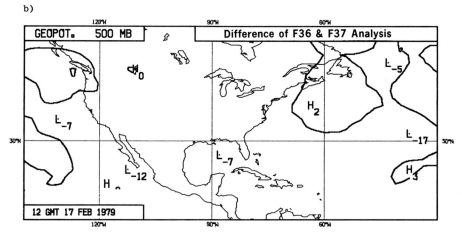

Fig. 9. (a) 500mb height analysis (F37) for 12 GMT 17 February 1979. In this analysis no satellite wind observations were used.

(b) Difference of analysis (F36), in which those observations were used, with the analysis shown in (a). Numbers in m

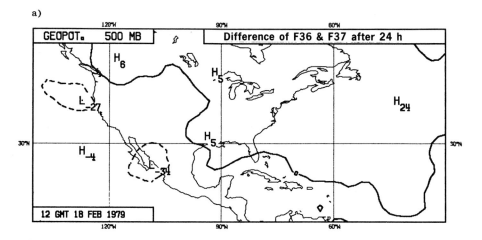

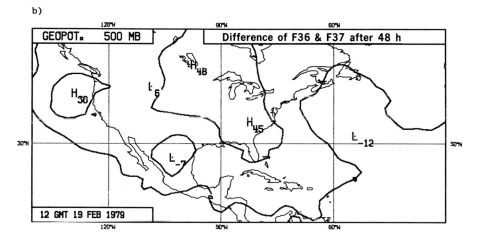

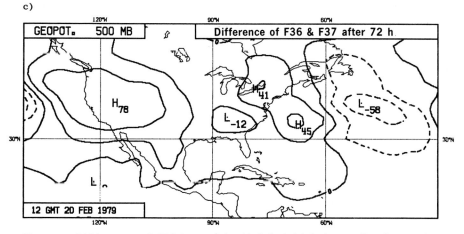

Fig. 10. Difference of 500mb geopotential height between the forecasts made from the analyses of Fig. 10 after 24h (a), 48h (b) and 72h (c). Numbers in m. Contours for strictly negative values have been dashed

a)

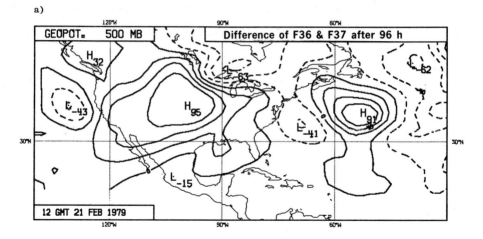

b)

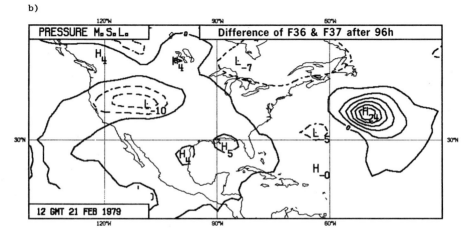

Fig. 11. Difference of 500mb geopotential height (m) (a), and mean sea level pressure (mb) (b), between the forecasts made from the analyses of Fig. 10 after 96 hours. Strictly negative contours have been dashed

those observations. The development of the difference between the two 500 mbar height analysis is shown in the figures. The analyses differ by at most 12 m (west of Mexico), but after four days the difference has amplified to 91 m. The surface pressure difference (Fig. 11b) at the position of the "President's-day storm" (40°N,50°W) is then 24 mbar. As might be expected, in such a sensitive situation, neither forecast is very good. Both underpredict the development of the low pressure system.

Gustafsson and Pailleux relate the explosive growth of the original difference to the vertical interpolation between pressure and sigma coordinates. The vertical interpolation interacted with cloud parameterization and radiative cooling in the model to produce different static stabilities in the two analyses. It was this change in static stability rather than the change in height fields that the forecast sensitivity originated from.

4.3 Hurrican Orchid

Originally in the ECMWF data assimilation scheme no correct distinction was made between real and virtual temperature when using the hydrostatic equation to transform the analysed variable height to the model variable temperature. In general the difference between the two temperatures is small, except in the humid tropical boundary layer, which plays a crucial role in the development of tropical hurricanes. Indeed, after the correction of the treatment of real and virtual temperatures, the hurricane Orchid was well predicted in a four day forecast from 4 September 1980, 1200 GMT (Fig. 12a), whereas it was hardly seen in the original forecast (Fig. 12b). The ECMWF verifying analysis is shown in Fig. 12c. This analysis has been obtained in a data assimilation run that included the analysis of Fig. 1c, 24 hours earlier. Had at that time the EREBO TEMP report been used (as in Fig. 1b) instead of the SYNOP report, the analysis at 8 September 1200 GMT would have shown a centre pressure of 998 mbar. Independent reports suggest that the

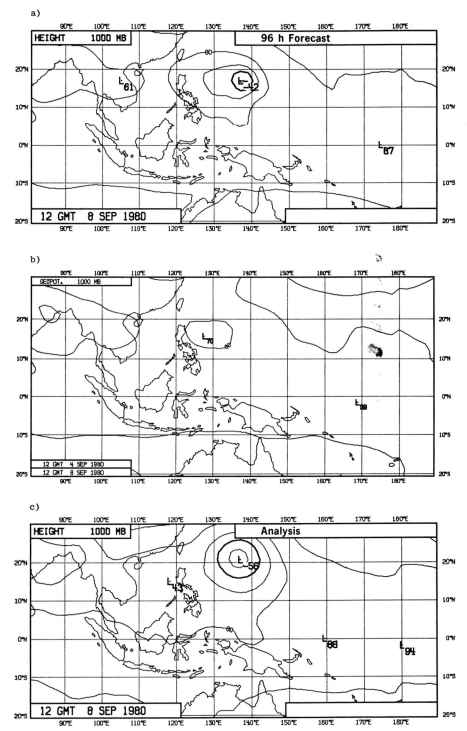

Fig. 12. Two four-day forecasts of hurricane Orchid valid for 12 GMT 8 September 1980.
  a) Correct treatment of humidity in the hydrostatic equation.
  b) Use of real instead of virtual temperature in the hydrostatic equation.
  c) Verifying analysis.
  Contour interval is 40 m

centre pressure was 985 mbar on 8 September, 1200 GMT.

4.4 Development of initial disturbances through the 10-day forecast

For 7 September 1980, 1200 GMT, two analyses were made, one using the TEMP report from the ship with call sign EREBO (Fig. 1b) and one using the SHIP report (Fig. 1c) for the 1000 mbar height. The two analyses were the same in all other respects. A similar procedure was followed for 8 April 1981, 1200 GMT, the only difference between the two analyses now being the treatment of the surface pressure report of the ship with call sign GWAN (Fig. 3).

The difference between two corresponding analyses can be considered as an initial perturbation. The growth of that perturbation during a 10-day forecast was studied by running forecasts from each analysis. Some properties of the perturbation are illustrated in Figs. 13 through 18 and in Table 1.

The disturbance survives initialization with a 1 mbar amplitude reduction in surface pressure (Table 1). With our sign convention the maximum disturbance is positive. The initial perturbation develops into a propagating and growing wavetrain (Figs. 13 and 14). There is a strong resemblance with figures shown by Simmons and Hoskins (1979). Other properties of the wavetrains correspond to their results as well, e.g. the zonal velocity of the forward fringe of the wavetrain is roughly 60% of the zonal wind speed at 200 mbar; new waves develop near the position of the initial disturbance; the systems have a vertical phase tilt; the trailing wave grows faster aloft than near the ground..

Simmons and Hoskins found also that the disturbance first required a few days before it really started to grow. Something similar is present in the 8 April 1981 case, but cannot be identified in the hurricane case. Fig. 15

Table 1. Some properties of the perturbations in the initial conditions for the two studied cases

	7 September 1980 (case: EREBO)	8 April 1981 case: GWAN)
Sign convention: Positive is:	accept TEMP, reject SHIP	accept SHIP
Max. difference in analyses:		
surface pressure (mbar)		
before initialization:	4	12
after initialization:	3	11
500 mb height, after initial.(m):	-4	56

Table 2. Precipitation from convective cloud, accumulated from the start of the forecast and surface pressure in the centre of hurricane Orchid in the 7 September 1980 case

Analysis uses EREBO report:	TEMP (998 mbar)		SHIP (992 mbar)	
	CONV. PREC.	MSLP	CONV. PREC.	MSLP
Time since forecast start (days)	(mm)	(mbar)	(mm)	(mbar)
0	0	1000	0	997
1	62	1001	92	997
2	82	1000	141	996
3	134	1001	253	997
4	176	998	255	993

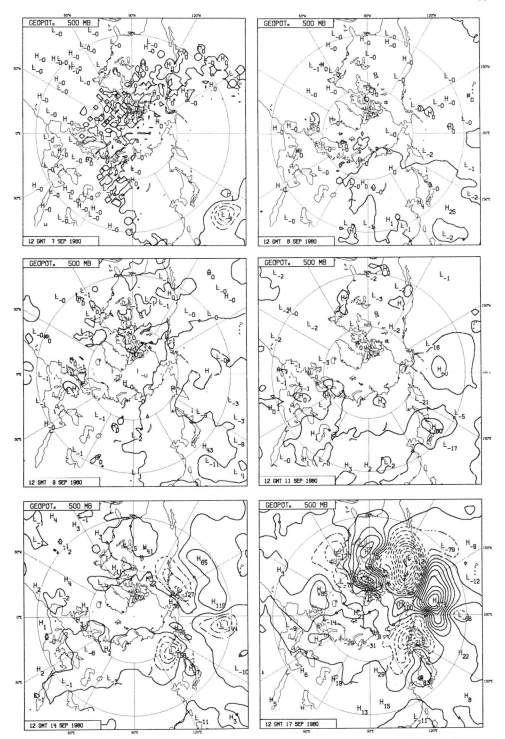

Fig. 13. Northern hemisphere 500 mb height difference between two analyses after initialisation (top left panel, contour interval 1 m) and between the subsequent forecasts after 1,2,4,7 and 10 days (contour interval 40 m). Full contours are for positive or zero values, dashed for negative.
The negligible amplitude noise, showing up in the zero contour and in the number of H and L symbols, is due to spectral truncation and initialisation.
Initial date: 7 September 1980

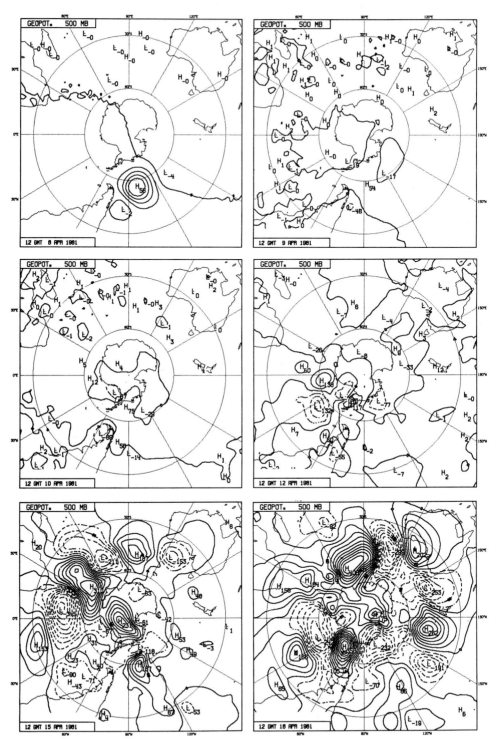

Fig. 14. As Fig. 14 for the southern hemisphere, initial date is 8 April 1981. The contour interval in the top left panel ('analysis') is 10 m, elsewhere it is 40 m

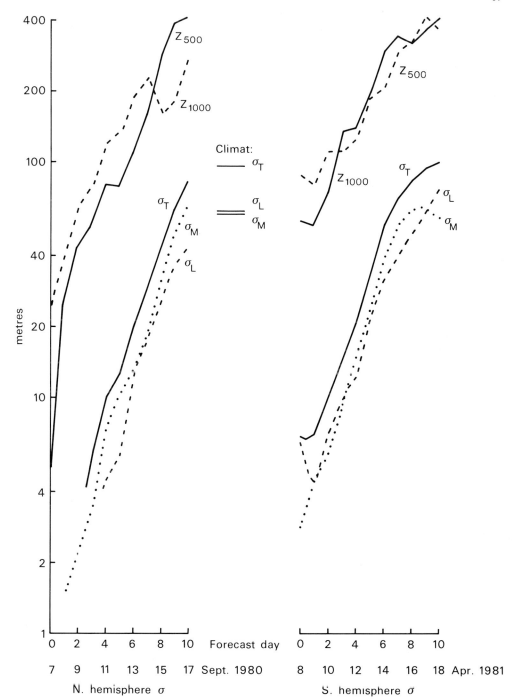

Fig. 15. Maximum amplitude of 1000 (Z_{1000}) and 500 mb height (Z_{500}) difference between the forecast pairs from 7 September 1980 and from 8 April 1981; and standard deviation of height difference between the forecasts over $20°-82.5°$ N and S respectively and 1000-200 mb in the zonal wavenumber bands: total (σ_T), 1-3 (σ_L) and 4-9 (σ_M). Also indicated is the September climatological standard deviation for the area $20°-82.5°$N, 1000-200 mb in the same zonal wavenumber bands

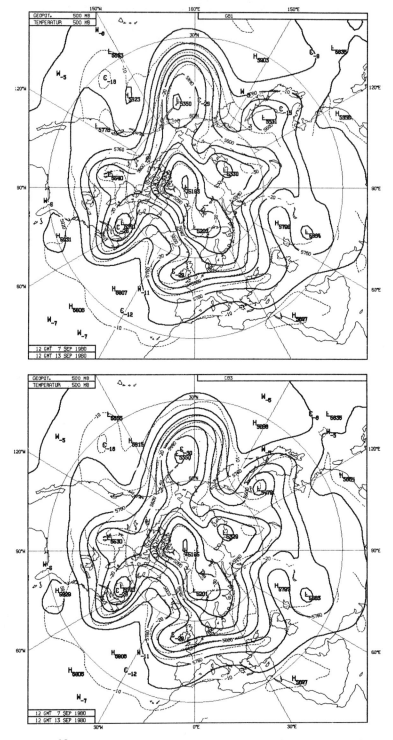

Fig. 16. Northern hemisphere 500 mb height (m, full lines) and temperature (°C, dashed lines) at day 6 in the two forecasts from 7 September 1980

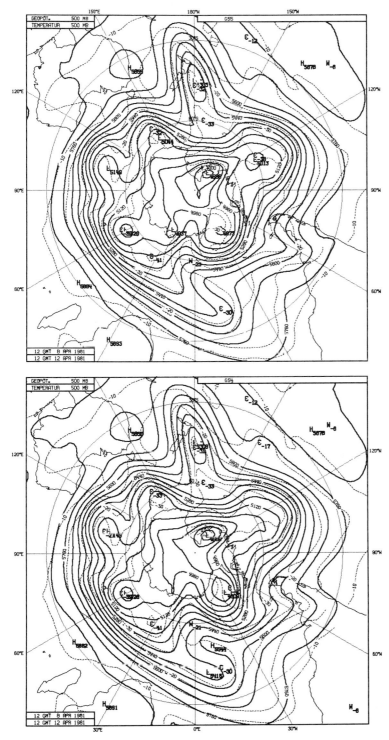

Fig. 17. Southern hemisphere 500 mb height (m, full lines) and temperature (°C, dashed lines) at day 4 in the two forecasts from 8 April 1981

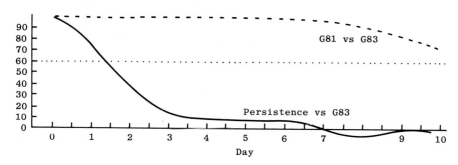

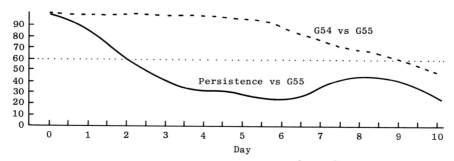

Fig. 18. Anomaly correlation of height over $20°$-$82.5°$, 1000-200 mb, between the two forecasts from 7 September 1980 (top, Northern Hemisphere) and those from 8 April 1981 (bottom, Southern Hemisphere). Heavy lines show the correlation of persistence with one of the forecasts

shows the growth rate of several error measures. The standard deviation between the geopotential heights in the two corresponding forecasts has been evaluated over the 1000 to 200 mbar standard pressure levels and over the latitude band 20° 82°.5 (north for 7 September 1980, south for 8 April 1981). The initial perturbation in the hurricane case is to a great extent outside this verification area, so that the initial growth cannot be inferred from this standard deviation. Fig. 15 shows that the maximum amplitude of the hurricane disturbance grows rapidly, also in the early stages, but both forecasts for this case do not develop the hurricane strongly (Table 2). (Phase shifts between the two forecasts account for the amplitude growth of the disturbance). Presumably the initial growth is due to latent heat release rather than baroclinic instability, and therefore the early stages of this case do not behave similarly to those shown by Simmons and Hoskins.

The standard deviation (Fig. 15) grows exponentially over a large range of values, with a doubling time of less than 2 days. This growth rate is the same in the two cases studied and rather independent of wavelength, as is shown by the separate display of the wavenumber groups 1-3 (σ_L) and 4-9 (σ_M). The maximum amplitude of the disturbance grows considerably slower, because this measure does not take the growth of the horizontal and vertical extent of the disturbance into account.

Although in both cases the forecasts differed considerably in the later stages of the forecast, the synoptic patterns were very much the same in the first four (8 April 1981) or six (7 September 1980) days (Figs. 16 and 17). The anomaly correlation verification score of the two forecasts from 7 September 1980 reached the 60% level with ECMWF analyses at day 6, and those from 8 April 1981 at day 4. These measures suggest that the atmosphere was predictable in the relevant hemisphere up to day 4 in the southern and up to day 6 in the northern hemisphere case.

In both cases the anomaly correlation between the two forecasts started only to deviate visibly from 1 by the time that the anomaly correlation between forecast and analysis went below 60%. An attempt has been made to obtain some feeling of how much of the more rapid decline of the latter correlation is due to the combined effect of initial errors and how much to forecast model errors. (Appendix B). A large number (estimates vary from twenty to forty, depending on the very tentative assumptions) of errors in the initial state of similar amplitude and growth rate would be required to decrease the 4 respectively 6 day correlation between two corresponding forecasts to that between a forecast and the verifying analysis, if it is assumed that the effects of the initial errors are uncorrelated and that the growth rate is the same for all initial errors. It is unlikely that the initial state contains so many large errors. There are of course many more much smaller errors, but every halving of initial amplitude requires a quadrupling of number of errors to produce the same effect. This suggests that the errors in the analysis certainly do not account for the forecast errors at the end of the integration, so that also errors must have been introduced during the integration. It is stressed, however, that the assumptions made to arrive at this conclusion are not necessarily correct due to the nonlinearity of the atmosphere and the model. A more quantitative estimate of these model errors relative to the analysis errors needs a deeper investigation.

Monte-Carlo type experiments with perturbations of the initial conditions suffered up to now from some problems. It has been difficult to make the initial disturbances survive initialization (Hollingsworth, 1979), but it appears that the cases presented here have overcome this problem. Another problem is the computational expense of Monte-Carlo experiments with a full model. If there is indeed sufficient similarity with the experiments of Simmons and Hoskins, a far cheaper model may be useable.

CONCLUSIONS AND OUTLOOK

At ECMWF, both during operational and experimental work, many cases have been encountered where the analyses and/or forecasts were very sensitive to certain parameters within the analysis scheme. Mostly the sensitivities are produced by discontinuities during data checking, but also the excitation of unstable waves plays a role occasionally. The correct value of many of the relevant parameters is not known yet; much of the work in the ECMWF analysis section is devoted to identifying and determining those parameters. Many problems, however, are due to inconsistencies or insufficiencies in the available observation systems. There is no hope that these latter problems will be solved in the near future. Our examples show that the quality of a medium range weather forecast may depend strongly on the quality of the initial analysis. With the present observational systems one should not expect that it is possible to produce forecasts with a consistently high quality. The average quality may be increased with suitable tuning of the data assimilation scheme.

The study of the development of an initial perturbation through a model integration may increase insight in the relative merits of a good analysis and a good model for medium range weather forecasting, and identify areas of atmospheric instability. A promising way of introducing the initial perturbation is by changing the quality control of one observational datum. The later development of such an initial perturbation in the complicated ECMWF model resembles that found in far simpler models.

Acknowledgement

The presented examples have been taken from experiments with the ECMWF data assimilation scheme. Most of the experiments have been run by others, and I am indebted to them for allowing me to use their results. I would like to thank particularly the members of the data assimilation group (J. Anderson, P. Lonnberg, D. Shaw, J. Van Maanen and W. Wergen) and of the FGGE group (N. Gustafsson, P. Kallberg, J. Pailleux and S. Uppala). I also thank A. Hollingsworth for his very useful suggestions and help.

Appendix A   ECMWF - GLOBAL FORECASTING SYSTEM, 15-level grid point model
(Horizontal Resolution $1.875°$ Lat/Lon)

ANALYSIS		PREDICTION
ϕ, u, v, q (for $p \geq 300$)		T, u, v, q

p(mb)		σ
10		$0.025(\sigma_1)$
20		0.077
30		0.132
50		0.193
70		0.260
100		0.334
150		0.415
200		0.500
250		0.589
300		0.678
400		0.765
500		0.845
700		0.914
850		0.967
1000	North	$0.996(\sigma_{15})$

Vertical and horizontal (latitude-longitude) grids and dispositions of variables in the analysis (left) and prediction (right) coordinate systems.

ANALYSIS

Method	3 dimensional multi-variate (15-analysis levels, see above)
Independent variables	λ, φ, p, t
Dependent variables	ϕ, u, v, q
Grid	Non-staggered, standard pressure levels
First guess	6 hour forecast (complete prediction model)
Data assimilation frequency	6 hour ($\pm$ 3 hour window)

INITIALISATION

Method	Non-linear normal mode, 5 vertical modes, adiabatic

PREDICTION

Independent variables	$\lambda, \varphi, \sigma, t$
Dependent variables	T, u, v, q, p_s
Grid	Staggered in the horizontal (Arakawa C-grid). Uniform horizontal (regular lat/lon). Non-uniform vertical spacing of levels (see above).
Finite difference scheme	Second order accuracy
Time-integration	Leapfrog, semi-implicit (Δt = 15 min) (time filter ν = 0.05)
Horizontal diffusion	Linear, fourth order (diffusion coefficient = $4.5 \cdot 10^{15} m^4 s^{-1}$)
Earth surface	Albedo, roughness, soil moisture, snow and ice specified geographically. Albedo, soil moisture and snow time dependent.
Orography	Averaged from high resolution (10') data set
Physical parameterisation	(i) Boundary eddy fluxes dependent on roughness length and local stability (Monin-Obukov)
	(ii) Free-atmosphere turbulent fluxes dependent on mixing length and Richardson number
	(iii) Kuo convection scheme
	(iv) Full interaction between radiation and clouds
	(v) Full hydrological cycle
	(vi) Computed land temperature, no diurnal cycle
	(vii) Climatological sea-surface temperature

APPENDIX B

EFFECT OF n UNCORRELATED ERRORS ON THE CORRELATION COEFFICIENT

In this Appendix the correlation coefficient between two fields ("forecast" and "analysis") will be calculated if the difference between forecast f and analysis a is the sum of n uncorrelated error fields ε_i (i=1...n):

$$f = a + \sum_{i=1}^{n} \varepsilon_i \qquad (B1)$$

$$\overline{\varepsilon_i \varepsilon_j} = \sigma_i^2 \text{ if } i=j$$

$$\qquad\qquad = 0 \text{ if } i \neq j \qquad (B2)$$

In (B2), σ_i^2 represents the variance of the error ε_i, and an overbar is used to denote averaging over the field. The notation $\tau_i = \sqrt{\sigma_i^2 / \overline{a^2}}$ will be used.

First, it will be assumed that analysis and error fields are uncorrelated:

$$\overline{a\,\varepsilon_i} = 0 \quad (i=1...n) \qquad (B3)$$

The correlation between forecast and analysis is then

$$c_{fa} = \frac{\overline{af}}{(\overline{a^2}\,\overline{f^2})^{\frac{1}{2}}} \qquad (B4)$$

with, according to (B1) and (B3):

$$\overline{af} = \overline{a^2} \qquad (B5)$$

and

$$\overline{f^2} = \overline{a^2} + \sum_{i=1}^{n} \sigma_i^2 = \overline{a^2}\,(1 + \sum_{i=1}^{n} \tau_i^2) \qquad (B6)$$

So:

$$c_{fa} = \frac{1}{(1 + \sum_{i=1}^{n} \tau_i^2)^{\frac{1}{2}}} \qquad (B7)$$

If all error fields have the same variance ($\tau_i^2 = \tau^2$ for all i), this becomes

$$c_{fa} = \frac{1}{(1 + n\tau^2)^{1/2}} \tag{B8}$$

Present forecast models do not increase the variance of the forecast field over that of the analysis field. A simple assumption, to replace (B3), accounts for this, namely:

$$\sum_i \overline{\epsilon_i a} = -\frac{1}{2} \sum_i \overline{\epsilon_i^2} = -\frac{1}{2} \overline{a^2} \sum_i \tau_i^2 \tag{B9}$$

With this correlation between analysis and error fields, (B6) is replaced by

$$\overline{f^2} = \overline{(a + \Sigma \epsilon_i)^2} = \overline{a^2} \tag{B10}$$

which indicates a constant forecast field variance.

Now (B5) and (B7) are replaced by, respectively

$$\overline{af} = \overline{a^2} - \frac{1}{2} \Sigma \sigma_i^2 \tag{B11}$$

and

$$c_{fa} = 1 - \frac{1}{2} \sum_{i=1}^{n} \tau_i^2 \tag{B12}$$

(B8) becomes

$$c_{fa} = 1 - \frac{1}{2} n \tau^2 \tag{B13}$$

In the text, two cases are presented. In both, the correlation between two forecasts is roughly 98% at the stage that the correlation between a forecast and the analysis is 60% (namely at day 4 for the 8 April 1981 case, and at day 6 for the 7 September 1980 case). If one of the forecasts was useable as a verifying analysis, both (B8) and (B13) would give $\tau^2 = 0.04$ for $c_{fa} = 0.98$ and n=1. The number n of error fields required to reduce c_{fa} to 60% is then about 40 (B8) or 20 (B13) (all error fields have been assumed to have the same variance). Because of the many and bold assumptions involved, these numbers should not be considered anything more than an order of magnitude estimate.

References

Gustafsson, N. and J.Pailleux 1981 On the quality of FGGE data and some remarks on the ECMWF data assimilation scheme. ECMWF Technical Memorandum 37, p 29.

Hollingsworth, A. 1979 An experiment in Monte-Carlo forecasting, In: Proceedings of the ECMWF workshop on Stochastic Dynamic Forecasting, Reading, 17-19 October.

Lorenc, A. 1981 A global three-dimensional multivariate statistical interpolation scheme. Mon.Wea.Rev., 109, 701-721.

Simmons, A. and B. Hoskins 1979 The downstream and upstream development of unstable baroclinic waves. J.Atmos.Sci., 36, 1239-1254.

Williamson,D. and C.Temperton 1981 Normal mode initialization for a multi-level grid-point model. Part II: nonlinear aspects. Mon.Wea.Rev., 109, 744-757.

Predictability of Time Averages:
Part I: Dynamical Predictability of Monthly Means[*]

J. SHUKLA

Abstract

We have attempted to determine the theoretical upper limit of dynamical predictability of monthly means for prescribed nonfluctuating external forcings. We have extended the concept of "classical" predictability, which primarily refers to the lack of predictability due mainly to the instabilities of synoptic-scale disturbances, to the predictability of time averages, which are determined by the predictability of low-frequency planetary waves. We have carried out 60-day integrations of a global general circulation model with nine different initial conditions but identical boundary conditions of sea surface temperature, snow, sea ice and soil moisture. Three of these initial conditions are the observed atmospheric conditions on 1 January of 1975, 1976 and 1977. The other six initial conditions are obtained by superimposing over the observed initial conditions a random perturbation comparable to the errors of observation. The root-mean-square (rms) error of random perturbation at all the grid points and all the model levels is 3 m s^{-1} in u and v components of wind. The rms vector wind error between the observed initial conditions is >15 m s^{-1}.

It is hypothesized that for a given averaging period, if the rms error among the time averages predicted from largely different initial conditions becomes comparable to the rms error among the time averages predicted from randomly perturbed initial conditions, the time averages are dynamically unpredictable. We have carried out the analysis of variance to compare the variability, among the three groups, due to largely different initial conditions, and within each group due to random perturbations.

It is found that the variances among the first 30-day means, predicted from largely different initial conditions, are significantly different from the variances due to random perturbations in the initial conditions, whereas the variances among 30-day means for days 31-60 are not distinguishable from the variances due to random initial perturbations. The 30-day means for days

[*] From the Journal of Atmospheric Sciences, vol. 30, No. 12, December 1981 with the kind permission of the American Meteorological Society

16-46 over certain areas are also significantly different from the variances due to random perturbations.

These results suggest that the evolution of long waves remains sufficiently predictable at least up to one month, and possibly up to 45 days, so that the combined effects of their own nonpredictability and their depredictabilization by synoptic-scale instabilities is not large enough to degrade the dynamical prediction of monthly means. The Northern Hemisphere appears to be more predictable than the Southern Hemisphere.

It is noteworthy that the lack of predictability for the second month is *not* because the model simulations relax to the same model state but because of very large departures in the simulated model states. This suggests that, with improvements in model resolution and physical parameterizations, there is potential for extending the predictability of time averages even beyond one month.

Here, we have examined only the dynamical predictability, because the boundary conditions are identical in all the integrations. Based on these results, and the possibility of additional predictability due to the influence of persistent anomalies of sea surface temperature, sea ice, snow and soil moisture, it is suggested that there is sufficient physical basis to undertake a systematic program to establish the feasibility of predicting monthly means by numerical integration of realistic dynamical models.

1. Introduction

The deterministic prediction of subsequent evolution of atmospheric states is limited to a few days due to the presence of dynamical instabilities and nonlinear interactions. Two predictions made from the same initial conditions, except for small and random differences in the initial state, begin to differ from each other (Lorenz, 1965; Charney et al. 1966; Smagorinsky, 1969). The rate and degree to which the two predictions diverge depends on the growth rates of the hydrodynamic instabilities, the nature of nonlinear interactions, and the structure of the differences between the two initial states. Since the observed state of the atmosphere, due to errors in observations and their interpolation to data-void areas, always contains some uncertainties, and since the formulation of the dynamical equations and parameterizations of the physical processes are only approximate, there is an upper limit on the range of deterministic prediction. This upper limit is mainly determined by the error growth rates associated with the instabilities of the mean flows with respect to the synoptic-scale disturbances and therefore it strongly depends on the structure of the initial conditions. Some initial conditions are more predictable than others. This upper limit may be referred to as the predictability limit for synoptic scales. Moreover, since the limits of predictability of synoptic scales are considered under fixed external forcing [including slowly varying sea surface temperature (SST), soil moisture, snow and ice, etc.], and since the evolution of the two initial states is determined completely by dynamical instabilities and their interactions, we propose to refer to it as the *dynamical predictability of synoptic scales*. In subsequent discussions we have distinguished between the *dynamical predictability* and *predictability due to external forcings*.

The dynamical predictability of synoptic scales is of interest for short-range prediction and considerable literature already exists on the different aspects of this problem. In this study we propose to investigate the predictability of space and time averages. We ask the following questions: While the detailed structure of instantaneous flow patterns cannot be predicted beyond a few days, is it possible that space and time averages can be predicted over longer averaging periods?

In an earlier paper, Charney (1960) had raised the question of predictability of space and time averages. It is perhaps appropriate to quote Charney:

"The crucial question is 'What is really remembered?' If, for example, the system only remembers a spatially or temporally averaged mean state, then it should at least be possible to predict this mean state. It seems to me that this is just the problem of long-range prediction."

The degree of dynamical predictability of space-time averages will naturally depend upon the space-time spectra of the atmospheric states and the nature of interaction among different scales. For example, if there were no stationary forcings at the earth's surface and if the day-to-day fluctuations were determined solely by the baroclinic instability of radiatively maintained zonal flows, the variability of time averages would mainly depend on the length of the averaging period and, therefore, the time averages will not be any more predictable than the amplitudes and phases of individual disturbances. In this case, most of the energy would be contained in the most unstable scales and larger scales will grow either through the cascade of energy from the fast growing unstable scales or due to their self amplification.

Recent studies by Charney and Devore (1979), Charney and Straus (1980) and Charney et al. (1981) have shown that for a given external forcing, thermally and orographically forced circulations can possess multiple equilibrium states and that some of these states are more stable than others. This suggests that in the presence of thermal and orographic forcings, self interaction and nonlinearities can be important mechanisms for fluctuations of longer periods. It is quite likely that large-amplitude synoptic instabilities play an important role in destabilizing the existing equilibria and thus the predictability of transitions from one equilibrium to another may not be any more promising than the predictability of intense cyclone waves.

Figure 1a shows the space-time spectra for the observed geopotential height field at 500 mb for 15 winter seasons. It is seen that most of the variance is contained in the planetary-scale (wavenumbers 1-4) low-frequency (10-90 day) components of the circulation. Figure 1b shows the variances in different wavenumber and frequency domains for 15 different winter seasons. Most of the interannual variability is contributed by the low-frequency planetary waves. Since the variability of space-time averages is mainly dominated by the planetary-scale low-frequency components, it can be anticipated that the prospects of predictability of space-time averages may not be as hopeless as those of the synoptic scales.

In the earlier classical studies of deterministic predictability, the theoretical upper limit was determined by the rate of growth and the magnitude of error between the two model evolutions for which the initial conditions differed by only a small random perturbation. For a study of the predictability of time averages, a more appropriate question would be: If numerical predictions are made from initial conditions which are as different as two randomly chosen years (for example, as different as observations on the same calendar date for different years), how long would it take before the time-averaged predictions become indistinguishable from the predictions made with small random perturbations in the initial conditions? The differences between two atmospheric states for the same calendar day in two different years can be very large because each dynamical state has evolved through complex nonlinear interactions under the influence of different boundary forcings, and therefore the amplitudes and phases of the main energy-bearing planetary waves will be much more different than those which can be expected by adding a small random perturbation. Thus it is natural to ask: Is it possible that a given configuration of planetary waves remembers itself much longer than the limit of deterministic prediction, which mainly refers to the predictability of synoptic scales?

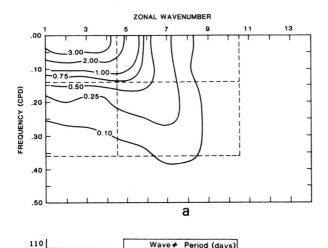

Fig. 1a. Wavenumber-frequency decomposition of transient variances of 500 mb geopotential height field along 50°N for 14 winter seasons

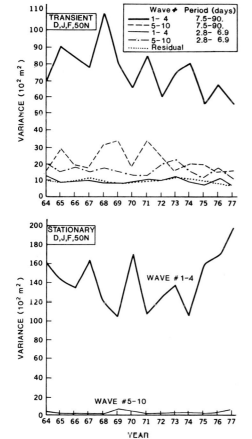

Fig. 1b. Interannual variability of transient variances (upper panel) and stationary variances (lower panel) of 500 mb geopotential height field along 50°N

The main difference between the previous studies and this study is that the earlier studies examined the growth rate of the errors, and the upper limit of synoptic-scale predictability was determined by the dominant instabilities. In the present study, we have carried out the analysis of variance to compare the variability among very different initial conditions and among randomly perturbed initial conditions. The limit of predictability is not determined solely by the growth rates of synoptic-scale instabilities, but by the relative magnitudes of the deterministic growth rates of the planetary waves and the degradation of planetary waves by the rapidly amplifying synoptic-scale instabilities. In other words, the lack of predictability of a high-frequency small-scale system itself is not very significant because its effect will be minimized due to averaging; however, its effects on the longer waves are important. The earlier studies of predictability considered the rate of error growth under fixed boundary conditions, i.e., they examined only the dynamical predictability of the initial conditions. There is the possibility that the predictability can be different for different boundary forcings. For example, time averages may be more predictable for certain anomalous structures of sea surface temperature, soil moisture, sea ice and snow.

The present study and the earlier studies share a common deficiency: both introduce only a random perturbation in the initial conditions. There is no evidence that the errors of observations are randomly distributed. Observational errors have systematic biases over the data-void areas (*viz.*, oceans) and it is not clear that the growth rates of systematic errors will be the same as the growth rates of random errors. However, just as the earlier predictability studies with random perturbations were intended mainly to illustrate the theoretical upper limit of synoptic-scale predictability, the present study also is an idealized study to establish the theoretical upper limit of predictability of time averages.

2. Mechanisms for the Interannual Variability of Monthly Means

Let us assume that $\Psi(\mathbf{X},t)$ represents the state of the atmosphere at any time t (schematic Fig.2), where $\mathbf{X}$ is the three-dimensional space vector.

At the earth's surface, which is the lower boundary of the atmosphere, sea surface temperature, sea ice and snow, soil moisture and vegetation, etc., act like slowly varying external forcings. Although solar variability is the only forcing truly external to the atmosphere, since the time scale

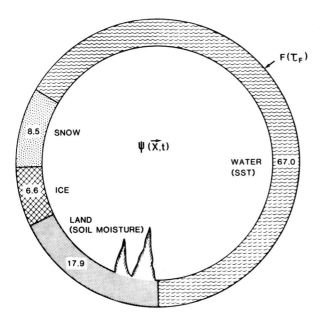

Fig.2. Schematic illustration of the roles of internal dynamics and slowly varying boundary conditions

of the change of such parameters as large-scale sea surface temperature, soil moisture and sea ice cover, etc., is much larger than the time scale of synoptic-scale instabilities, we would refer to these slowly varying boundary conditions as external forcings to the atmospheric system described by $\Psi(X,t)$. The time evolution of Ψ can now be treated as the combined effects of the initial conditions (because they will determine the nature of instabilities and their interactions) and the effects of external forcing as defined by the boundary conditions. There are three distinct time scales which enter any consideration of the predictability of time averages. First is the time scale of day-to-day fluctuations of Ψ (to be denoted by τ_Ψ) due to inherent instability and nonlinearity of the system, and this determines the level of uncertainty in estimating a time average (Leith, 1973), second is the time scale of external forcing (to be denoted by τ_F); and third is the averaging time (T) for which we wish to examine the predictability. For the particular case of predictability of monthly means, it can be stated that $\tau_\Psi < T < \tau_F$, because τ_Ψ for the atmosphere is a few days and τ_F for large-scale sea surface temperature and sea ice anomalies is a few months (we do not have a good observational basis to estimate τ_F for soil moisture). Since these boundary conditions change slowly, they can be assumed to be constant (or prescribed) for a period of one month. However, even under fixed external forcings, internal dynamics can change the evolution of initial states to such an extent that time averages will be different from each other. The mechanisms responsible for the interannual variability of monthly means can be broadly categorized as follows:

(i) *Internal dynamics*. Due to combined effects of instabilities, nonlinear interactions, thermal and orographic forcings and fluctuating zonal winds, tropical-extratropical interactions, etc. (Since orography and land-sea contrast at the earth's surface is fixed, thermally and topographically forced motions and their interactions are considered as a part of the internal dynamics).

(ii) *Boundary forcings*. Due to fluctuations of sea surface temperature, sea ice/snow, soil moisture and other slowly varying boundary conditions and their effects on the amplitudes and phases of planetary waves which in turn may determine the tracks and intensity of cyclone-scale disturbances. Fluctuations in solar or other extraterrestrial energy sources are not considered in the present study.

An understanding of the relative contributions of the two factors is essential to determine the predictability of climate in general and of monthly means in particular. We recognize that the dynamical properties of the small-scale instabilities can be very different under extremely different boundary conditions. However, the observed anomalies in the boundary conditions are not found to be so large that they could change the basic dynamics of the synoptic disturbances. The changes in the boundary conditions can affect the amplitudes and phases of the planetary waves in extratropical latitudes and large-scale Hadley and Walker circulations in low latitudes. Changes in the structure and persistence of the midlatitude planetary waves, in turn, can change the frequency, intensity and propagation properties of synoptic-scale instabilities. Similarly, in low latitudes, SST and soil moisture anomalies of moderate magnitude can change the growth and amplitude of synoptic-scale tropical disturbances, whereas anomalies of large magnitude can drastically alter the geographical locations of convergence zones and precipitation. Anomalous heat sources in low latitudes can also produce, under favorable conditions, large responses in middle latitudes. In this paper, however, we propose to study the predictability of monthly means under fixed boundary conditions. This we would refer to as the *dynamical predictability* of monthly means. Examination of the role of slowly varying boundary conditions in determining the changes of monthly means is a topic for separate investigation.

General circulation models allow us to examine the dynamical predictability and boundary-forced predictability separately. It is not possible to address these questions by analysis of observed data alone because the observations reflect the combined effects of internal dynamics and fluctuating boundary conditions. Even in the absence of fluctuating external forcings, internal dynamics can generate interannual variations of monthly and seasonal means. Sometimes, this component of the variability has been referred to as the "climate noise" (Leith, 1975; Straus and Halem, 1981). This terminology can be misleading if the word "noise" refers implicitly to those components of the circulation which are considered to be unpredictable. Since it has not yet been established that the changes due to internal dynamics are unpredictable, it is not appropriate to refer to the variability due to internal dynamics as climate noise. We propose to distinguish between the fast growing synoptic instabilities and slowly varying planetary scales which possess different predictability characteristics. Lack of synoptic-scale predictability beyond two weeks is not sufficient to assume that certain statistical properties (e.g., monthly means) of the internal dynamics are not predictable. It is quite likely that some components of the atmospheric flows may retain a certain degree of nonlinear memory for a period beyond the limits of deterministic prediction for synoptic scales and therefore a nonlinear dynamical prediction scheme could predict the evolution of large-scale components over a period longer than that for synoptic scales.

3. Numerical Experiments

We have carried out 60-day integrations of the GLAS climate model starting from three different initial conditions corresponding to 1 January of 1975, 1976 and 1977. These are referred to as the control runs. The differences among the initial conditions for 1 January of different years are larger than those due to errors of observations because they reflect the multitude of effects of varying boundary conditions and dynamical interactions during each preceding year. Amplitudes and phases of planetary waves were significantly different from each other. The rms vector wind differences at 500 mb for observed initial conditions over Northern Hemisphere were 16.9 m s^{-1} between 1975 and 1976, 17.6 m s^{-1} between 1976 and 1977, and 18.5 m s^{-1} between 1975 and 1977; the differences were ~10 m s^{-1} in the lowest tropospheric levels and ~20 m s^{-1} in the upper troposphere. The rms differences for the initial conditions of 500 mb geopotential height between 30 and 70°N were 160.4 m between 1975 and 1976, 187.7 m between 1976 and 1977, and 186.7 m between 1975 and 1977. In comparison, the rms difference between maps from two randomly chosen years, calculated from 15 years of daily values, was found to be 178.8 m.

Each of these initial conditions was then randomly perturbed such that the spatial structure of the random perturbations in u and v components at all the nine levels of the model had a Gaussian distribution with zero mean and standard deviation of 3 m s^{-1}. The amplitudes of the random perturbation in u and v were not allowed to exceed 12 m s^{-1} at any grid point. These are referred to as the perturbation runs. For each control run there are several perturbation runs. The general circulation model used in the present study has been described by Halem et al. (1980).

Since all the integrations are made with fixed boundary conditions, variability among the predictions of monthly means made from the control runs gives a measure of the long-range memory of the initial conditions characterized by different configurations of the planetary waves. Variability among different perturbation runs for the same control run gives a measure of the degradation of the predictability of those initial conditions. If for a given averaging period T, the variability among the control runs is not statistically different from the variability among the perturbation runs, we can conclude that there is no dynamical predictability for the time averages over period T. It should be pointed out, however, that this statement applies only to the model being used for the study and not necessarily to the real atmosphere. However, since ultimately one or the other model has to be used for any prediction, lack of predictability for the models will imply our inability to make actual predictions of the atmosphere. Absence of *dynamical predictability* would not imply, however, the absence of boundary-forced predictability due to anomalous structures of sea surface temperature, soil moisture and sea ice, etc.

Table 1 gives a summary of the numerical integrations carried out for this study. C_{11}, C_{21}, and C_{31} refer to the three 60-day control runs made from the initial conditions of 1 Januar 1975, 1976 and 1977, respectively. C_{12}, C_{13}, are the perturbation runs for which the initial conditions of C_{11} were randomly perturbed. Likewise, C_{22}, C_{23} and C_{24} are the perturbation runs for the control run C_{21}, and C_{32} is the perturbation run for the control run C_{31}. The statistical properties of the random perturbations were the same in all the perturbation cases; however, the actual values at any grid point were different in each case.

Table 1. Schematic summary of the control and the perturbation runs (C_{ji})

	Initial condition	1 Jan 1975 $j = 1$	1 Jan 1976 $j = 2$	1 Jan 1977 $j = 3$
Control	$i = 1$	C_{11}	C_{21}	C_{31}
Random perturbation	$i = 2$	C_{12}	C_{22}	C_{32}
Random perturbation	$i = 3$	C_{13}	C_{23}	
Random perturbation	$i = 4$		C_{24}	

4. Predictability of Planetary and Synoptic Scales

We have examined the scale dependence of the theoretical upper limit of deterministic predictability. The errors for planetary waves and synoptic-scale waves are very different. We show that the planetary waves which have the largest contribution to monthly means have a much longer predictability time compared to the synoptic-scale waves.

The 500 mb geopotential height (ϕ) for each day and each latitude for control (p) and perturbation (c) runs can be expressed as

$$\phi^c(\lambda,t) = \phi_0^c + \sum A_k^c \cos\theta_k + B_k^c \sin\theta_k ,$$

$$\phi^p(\lambda,t) = \phi_0^p + \sum A_k^p \cos\theta_k + B_k^p \sin\theta_k ,$$

where λ is the longitude and θ_k the phase for the wavenumber k. The mean value of 72 grid points along the latitude circle is denoted by $k = 0$. The total mean-square error E along the latitude circle can be expressed as

$$\{E(t)\}^2 = \frac{1}{2\pi} \int_0^{2\pi} [\phi^c(\lambda,t) - \phi^p(\lambda,t)]^2 d\lambda$$

$$= (A_0^c - A_0^p)^2 + \frac{1}{2} \sum_{k=1}^{36} (A_k^c - A_k^p)^2 + (B_k^c - B_k^p)^2 ,$$

where A_k and B_k are functions of time.

The right-hand side gives the contribution of each wavenumber to the total rms error. We have examined the errors, as defined below, in the planetary scales (wavenumbers 0-4), synoptic scales (wavenumbers 5-12) and short scales (wavenumbers 13-36) separately:

Planetary-scale error =

$$\left[(A_0^c - A_0^p)^2 + \sum_{k=1}^{4} (A_k^c - A_k^p)^2 + (B_k^c - B_k^p)^2 \right]^{1/2}$$

Synoptic-scale error =

$$\left[\sum_{k=5}^{12} (A_k^c - A_k^p)^2 + (B_k^c - B_k^p)^2 \right]^{1/2}$$

Short-scale error =

$$\left[\sum_{k=13}^{36} (A_k^c - A_k^p)^2 + (B_k^c - B_k^p)^2 \right]^{1/2} .$$

We have calculated the daily values of errors for the six pairs (see Table 1) of control and perturbation runs: (C_{11}, C_{12}), (C_{11}, C_{13}), (C_{21}, C_{22}), (C_{21}, C_{23}), (C_{21}, C_{24}), (C_{31}, C_{32}). Figures 3a and 3b show the mean of the six error values for 500 mb geopotential height for wavenumbers 0-4 and 5-12 respectively, averaged over the latitude belt 40-60°N. The vertical bars on each curve give the standard deviation among six error values. Mean and standard deviations of the persistence error in the corresponding wavenumber range for the three control runs, C_{11}, C_{21}, C_{31} also are shown on each figure. Errors in wavenumbers 13-36 are very small and are not shown here.

The differences in the two figures are rather remarkable. If the cross-over point between the random error growth curve and the persistence error curve is considered to be the theoretical upper limit of the deterministic predictability, the synoptic scales are found to lose complete predictability after two weeks. The persistence error is a measure of error between two randomly chosen charts and is $\sqrt{2}$ times larger than the rms of daily fluctuations (climatology error). The planetary scales, on the other hand, seem to show theoretical predictability even beyond one month. This further suggests that the space and time averages, which are mainly determined by the planetary-scale motions, have a potential for predictability at least up to or beyond one month.

We also have examined the day-to-day changes of sea level pressure and 500 mb temperature for control run and perturbation runs. Figures 4a-4f show the plots of daily sea level pressure and temperature averaged over some of the areas shown in Fig.7. It can be seen that the initial conditions do not persist in the course of integration and even large spatial averages show large day-to-day fluctuations. During the first month, although the

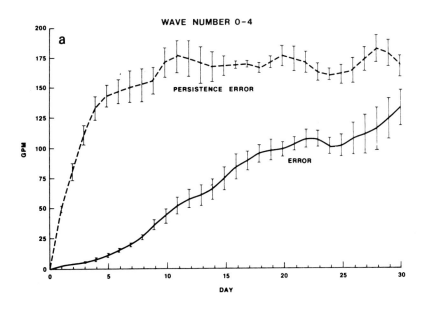
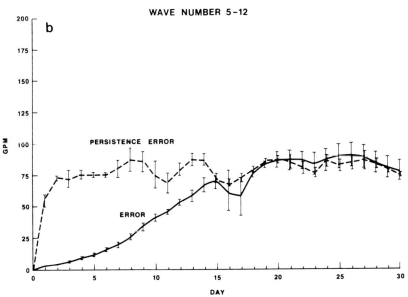

Fig.3a. Root mean square error, averaged for six pairs of control and perturbation runs and averaged for latitude belt 40-60°N for 500 mb geopotential height (gpm), for wavenumbers 0-4. Dashed line is the persistence error averaged for the three control runs. Vertical bars denote the standard deviation of the error values

Fig.3.b. As in Fig.3a except for wavenumbers 5-12

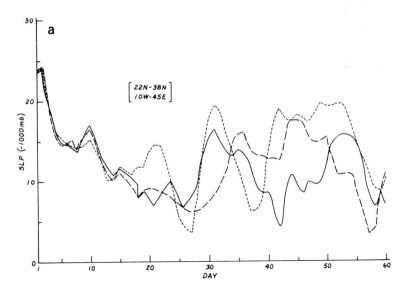

Fig.4a. Daily values of sea level pressure (mb) averaged for grid points between 22 and 38°N, 10°W and 45°E, for control run (solid line) and two perturbation runs (dashed lines) for initial conditions of 1 January 1975

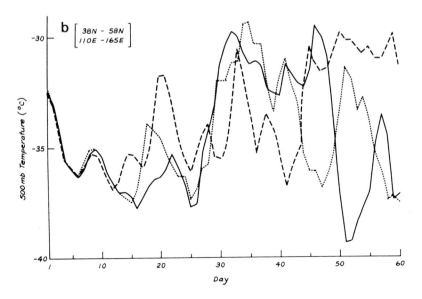

Fig.4b. Daily values of 500 mb temperature (°C) averaged for grid points between 38 and 58°N, 110 and 165°E, for control run (solid line) and two perturbation runs (dashed lines) for initial conditions of 1 January 1976

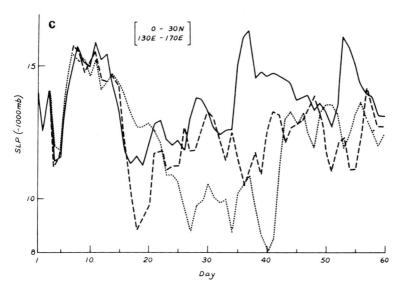

Fig.4c. Daily values of sea level pressure (mb) averaged for grid points between 0 and 30°N, 130 and 170°E, for control run (solid line) and two perturbation runs (dashed lines) for initial condition of 1 January 1975

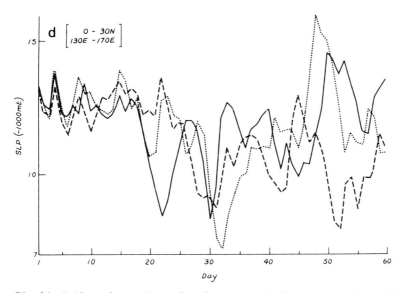

Fig.4d. Daily values of sea level pressure (mb) averaged for grid points between 0 and 30°N, 130 and 170°E, for control run (solid line) and two perturbation runs (dashed lines) for initial conditions of 1 January 1976

124

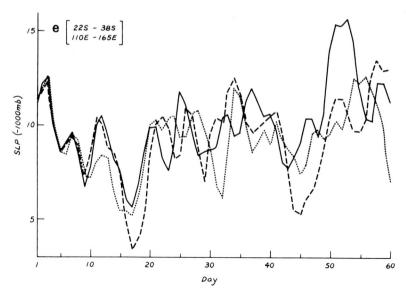

Fig.4e. Daily values of sea level pressure (mb) averaged for grid points between 22 and 38°S and 110 and 165°E, for control run (solid line) and two perturbation runs (dashed lines) for initial conditions of 1 January 1975

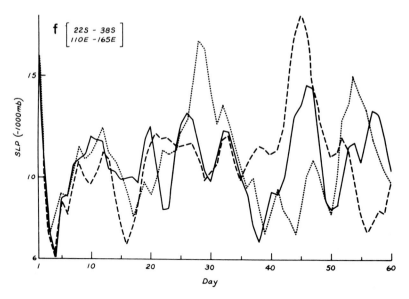

Fig.4f. Daily values of sea level pressure (mb) averaged for grid points between 22 and 38°S and 110 and 165°E, for control run (solid line) and two perturbation runs (dashed lines) for initial conditions of 1 January 1976

day-to-day fluctuations of either the control or the perturbation runs are large, the differences between the control run and the random perturbation run are not as large. However, for the second month, the departures between the control and the perturbation runs are so large that even the monthly means are indistinguishable from the monthly means for a completely differential initial condition.

Relatively small differences between spatially averaged atmospheric variables for the first 20-25 days of control and perturbation runs would suggest that the time averages for the first month should be predictable. We have shown this to be the case by a more systematic analysis of variance of model simulations described in the following sections.

5. Results

Figures 5a-5c show the plots of monthly mean sea level pressure for January (day 1-31) for the model runs (C_{11},C_{12}),(C_{21},C_{22}) and (C_{31},C_{32}), respectively. The upper panel shows the monthly mean for the control run and the lower panel shows the monthly mean for the random perturbations over that control run. To reduce the number of figures and to emphasize the large changes in winter, we have only shown the Northern Hemisphere maps. For January, any control run is much more similar to its own perturbation runs than to any other control run or any other perturbation run. The upper and lower panels of Figs.5a-5c have large similarities compared to any two upper panels or any two lower panels. For example, both the upper and the lower panels of Fig.5c show a very deep Aleutian low which is neither as strong nor located at the same place in other maps for January. The same is true for other major circulation features. This suggests that the random perturbations in the initial conditions have not changed the 31-day evolution of the flow so drastically that the monthly means may look very different.

The results are very different for the month of February (days 32-60), which are shown in Figs.6a-6c. Although some of the large-scale features retain their general configuration in the upper and lower panels, the displacements in the centers of highs and lows is large enough to give very significant qualitative differences between the upper and the lower panels. The differences between the upper and the lower panels are comparable to the differences between any two figures. In the following section, we present a more quantitative description of the differences between the control and the perturbation runs.

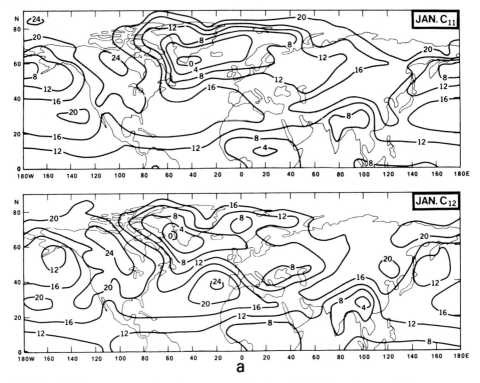

Fig.5a. 31-day mean sea level pressure for days 1-31 for the control run from the initial conditions of 1 January 1975 (upper panel: JAN C_{11}) and its perturbation run (lower panel: JAN C_{12})

a. *Space and Time Averaging*

We have examined the predictability of monthly means for several space averaging domains. Let us first ask a simple question: Why do we examine monthly means rather than examining 23-day or 39-day means? We have not carried out, and we are not aware of, any study which examines the most appropriate averaging periods for time-averaged predictions. Naturally the answer would depend on the frequency spectra of the different components of the flow. Since the space and time scales of atmospheric motions are dynamically related, it also should depend on the appropriate combination of space and time averaging domains. Our choice of monthly time period is based simply on a qualitative reasoning that it is more than two weeks, which is the upper limit of deterministic prediction, and less than a season which is an appropriate time scale for the external solar forcing. Moreover, for social consumption, forecasts are normally given in terms of monthly and seasonal means.

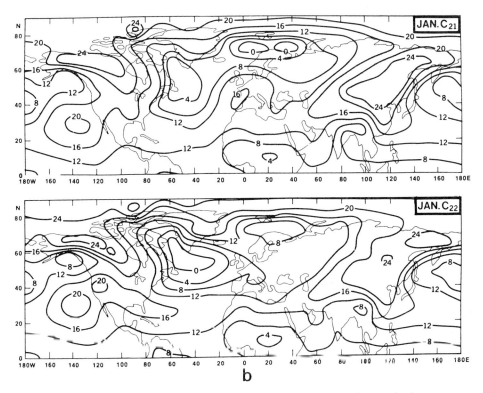

Fig.5b. 31-day mean sea level pressure for days 1-31 for the control run from the initial conditions of 1 Januar 1976 (upper panel: JAN C_{21}) and its perturbation run (lower panel: JAN C_{22})

Similarly, there is arbitrariness in choosing the space scales for averaging. From an examination of the past data over United States one can see that the space scales of the monthly mean anomalies for the large-scale dynamical variables (e.g., pressure, temperature, wind, etc.) is always larger than 500 × 500 km, which is the resolution of the model, but not as large as the whole of continental United States. In fact, typically, different quadrants of United States show different "signs" of monthly and seasonal anomaly field. It would, therefore, be meaningless to predict a monthly mean for the whole United States. In this study we have examined the monthly means averaged over different areas shown in Fig.7. While choosing the averaging area we have tried to isolate the oceanic and continental areas, the monsoonal and nonmonsoonal areas and the area of active midlatitude cyclogenesis.

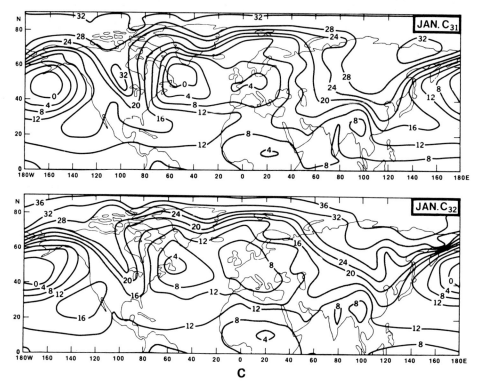

Fig.5c. 31-day mean sea level pressure for days 1-31 for the control run from the initial conditions of 1 January 1977 (upper panel: JAN C_{31}) and its perturbation run (lower panel: JAN C_{32})

b. *Analysis of Variance*

We have carried out the analysis of variance (F test) to determine the statistical significance of the differences in the variances among different control runs and among different perturbation runs. This technique is described in statistics text books (e.g., Hays, 1963).

In Table 1 the three boxes labeled as j = 1, 2, 3 refer to the three different initial conditions. For each box j, index i refers to the perturbed runs made with the initial condition for box j. If M_{ji} denotes the monthly mean value of any variable for the model run C_{ji}, and J is the number of boxes, the expression for F can be written as

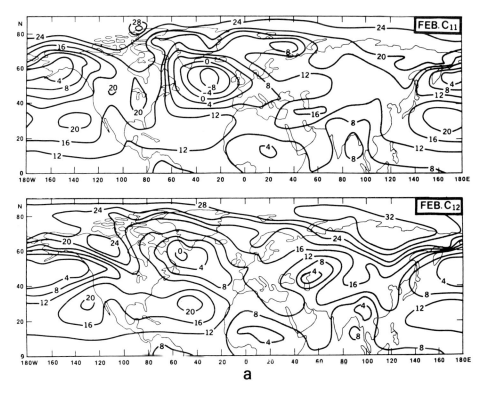

Fig.6a. 29-day mean sea level pressure for days 32-60 for the control run from the initial conditions of 1 Januar 1975 (upper panel: FEB C_{11}) and its perturbation run (lower panel: FEB C_{12})

$$F(\nu_1,\nu_2) = \frac{\sum_j \frac{\left(\sum_i M_{ji}\right)^2}{n_j} - \frac{\left(\sum_j \sum_i M_{ji}\right)^2}{N}}{\left(\sum_j \sum_i M_{ji}\right)^2 - \sum_j \frac{\left(\sum_i M_{ji}\right)^2}{n_j}} \cdot \frac{(N-J)}{(J-1)}$$

$$= \frac{F_1}{F_2} \frac{(N-J)}{(J-1)} .$$

In the present case (see Table 1) $n_1 = 3$, $n_2 = 4$, $n_3 = 2$, $N = n_1 + n_2 + n_3 = 9$, $\nu_1 = J - 1 = 2$, $\nu_2 = N - J = 6$.

The numerator (F_1) is a measure of variability among the monthly means of largely different initial conditions and the denominator (F_2) is a measure of variability among the monthly means for randomly perturbed initial

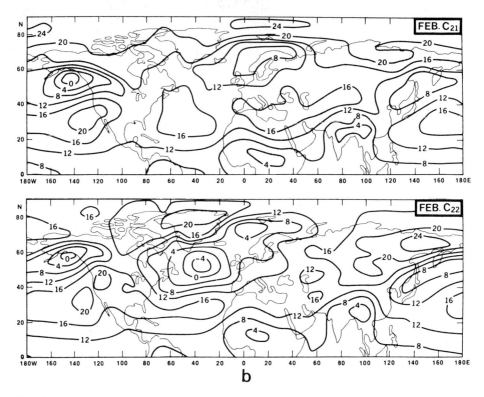

Fig.6b. 29-day mean sea level pressure for days 32-60 for the control run from the initial conditions of 1 January 1976 (upper panel: FEB C_{21}) and its perturbation run (lower panel: FEB C_{22})

conditions. By carrying out an analysis of variance, we are examining the significance of the fluctuations of the monthly means and not the monthly means themselves.

The algebraic mean of all the runs in any one box differs from the algebraic mean of all the runs in another box. The numerator F_1 is a measure of the variability among such box means, and the denominator F_2 is a measure of variability within the boxes. If the three different initial conditions are indeed very different, and if the random perturbations do not change them substantially, the numerator will be large and remain large, whereas the denominator will be small and remain small. However, if the random perturbations can produce large changes in the time averages, the denominator will increase and reduce the value of F. A reduction in the value of the numerator will also reduce the value of F. This will occur if the algebraic mean of all the runs in one box is not very different from the algebraic mean of all the runs in other boxes.

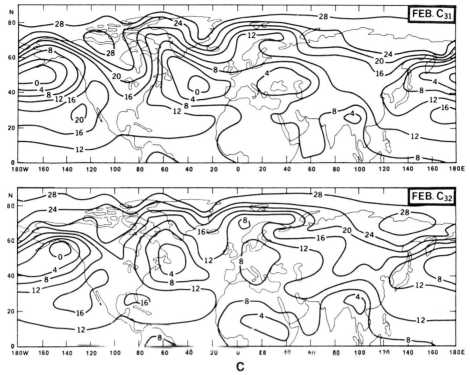

Fig.6c. 29-day mean sea level pressure for days 32-60 for the control run from the initial conditions of 1 January 1977 (upper panel: FEB C_{31}) and its perturbation run (lower panel: FEB C_{32})

The numerical value of F determines the level of significance for the differences between the variability among the boxes and the variability within the boxes. From F tables, $(\nu_1, \nu_2) = 5.1$ for 95% and 10.9 for 99% and 14.5 for 99.5% level of significance. Therefore, if the value of F, calculated from Eq. (1) for all the integrations in Table 1, is larger than 5.1, we can conclude that the variability among the monthly means of the control runs with very different initial conditions is larger (at the 95% significance level) than the variability among the monthly means due to random perturbations in the initial conditions.

For any initial condition, the day-to-day evolution of the flow will be different between the control run and its perturbation run. This will lead to different values of monthly means. The magnitude of this difference will depend, among other things, on the number of days the flow is allowed to evolve before calculating the monthly means. For example, the differences between the monthly means of the control run and the perturbation run for

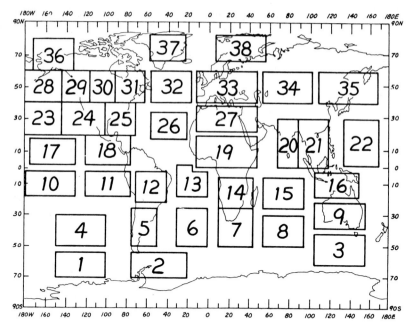

Fig.7. Locations and sizes for the 38 areas over which sea level pressure is averaged for the analysis of variance

days 31-60 will be different from those for days 1-30. Similarly, the differences between the monthly means of two control runs will be different from the differences between the monthly means of a control run and its perturbation run. The purpose of this analysis is to introduce a quantitative measure to examine these differences and compare them with each other. If for days 31-60, the variability among the control runs is not significantly larger than the variability among the perturbation runs, it will suggest that during a 60-day integration of the global GCM, the evolution of the flow was so modified by the presence of random errors in the initial condition that even a 30-day mean (for the last 30 days) was indistinguishable from a similar 30-day mean for very different initial conditions. Since the boundary conditions are identical for all the integrations, this will imply that the different initial conditions of the control runs had no bearing on the time averages for days 31-60.

In this study we have carried out 60-day numerical integrations for all the nine cases. We have first examined the predictability of time averages for days 1-31, 16-46 and 32-60. For convenience we refer to these as January (J), January-February (J/F) and February (F), respectively.

It is recognized that the observed initial conditions on any day are not unrelated to the observed boundary conditions for the same day and therefore certain inconsistencies might occur through using climatological mean boundary conditions. Presumably, this inconsistency is common to all the model runs with different initial conditions. The error growth rates were very similar for all the perturbation runs. Since we propose to study the dynamical predictability under identical boundary conditions, we had no better alternative than choosing the climatological mean boundary conditions. A more appropriate procedure would have been to choose the control initial conditions from a long (several years) GCM integration carried out with constant (or seasonally varying) boundary conditions. Due to limitations on available computer time, we could not carry out such long integrations. However, such long integrations already have been carried out by other modeling groups, (S. Manabe and M. Schlesinger, personal communication) and it may be useful to carry out similar studies with these models.

c. *Results of F calculation*

Figures 8a-8c give the F values for sea level pressure averaged over each area for the averaging periods of days 1-31, 16-46 and 32-60, respectively. Areas with F values of 5 or more are shaded. For the monthly means of days 1-31, 29 out of 38 areas show F values of 5 or more. The number of such significant areas drops down to 11 for monthly means of days 16-46 and drops further down to 5 for monthly means of days 32-60. Since two of these five areas did not show significance for averaging periods of days 16-46, it is reasonable to conclude that only three of the areas show significance for the averaging period of days 32-60. Since we are considering the significance level above 95%, two of 38 areas may be significant by chance. It can be concluded that there is a complete loss of dynamical predictability for averaging periods of days 32-60.

If the number of significant areas are segregated according to the hemisphere, it is seen that only three out of 18 areas in the Southern Hemisphere are significant above 95% for the averaging period of days 16-46. This suggests that the Northern Hemisphere monthly means are potentially more predictable than the Southern Hemispheric monthly means. It is quite likely that the presence of orographic and thermal forcings due to mountains and continental-oceanic heat sources in the Northern Hemisphere winter establishes planetary-scale motions which are sufficiently stable to allow a dynamical prediction for a longer range compared to the Southern Hemisphere.

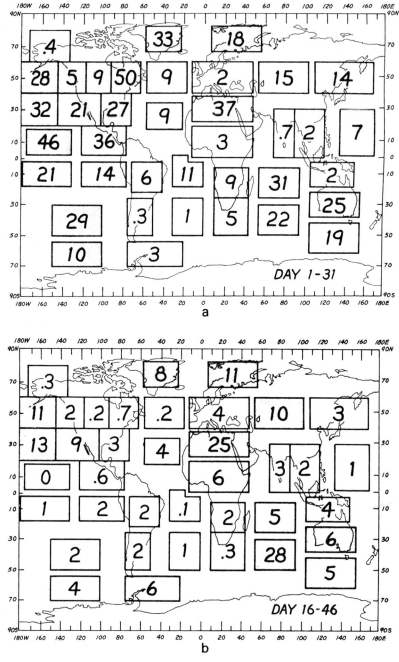

Figs.8a and b (Captions see opposite page)

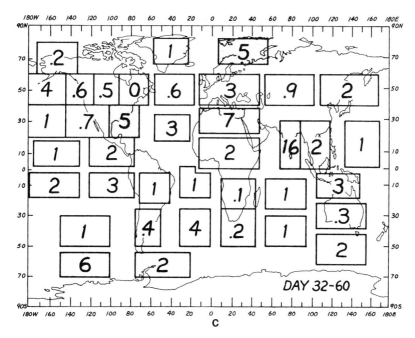

Fig.8c.

Fig.8a. F values for 31-day mean (days 1-31) sea level pressure averaged over different areas

Fig.8b. F values for 31-day mean (days 16-46) sea level pressure averaged over different areas

Fig.8c. F values for 29-day mean (days 32-60) sea level pressure averaged over different areas

It may be conjectured likewise that the prospects of longer term predictability for the Northern Hemispheric summer may not be as favorable.

We have examined the sensitivity of these results to the size of the averaging area. We have repeated the calculations of F for areas smaller and larger than those shown in Fig.7 but centered at the same 38 areas. We refer to the modified areas as Area-1, Area+1, Area+2, and Area+4, respectively. Area+n (Area-n) refers to an area which is increased (decreased) by n grid points on either side of the area shown in Fig.7. One grid length in latitude and longitude is 4 and 5°, respectively. Tables 2a-2c give the values of F for the averaging periods of days 1-31, 16-46 and 32-60, respectively. For January, the number of boxes significant at more than the 95% confidence level are maximum (-29) for the areas shown in Fig.7 and Area+1. There is a slight reduction in the number of significant areas for smaller and larger areas. The effect seems to be more clear for Southern Hemisphere

Table 2a. Values of F for January (days 1-31) for different sizes of the area of averaging

Region no.	(Area − 1)	Area (Fig.7)	(Area + 1)	(Area + 2)	(Area + 4)
1	9.0	10.3	14.9	19.6	8.9
2	3.7	3.4	3.1	3.4	3.6
3	25.6	18.6	17.4	19.6	39.3
4	40.3	29.0	22.2	12.4	1.9
5	0.1	0.2	0.3	0.2	0.2
6	1.4	0.9	0.5	0.3	0.0
7	5.7	5.1	4.1	2.5	1.3
8	23.9	22.2	19.5	16.0	8.4
9	16.0	25.4	31.1	32.9	30.5
10	32.8	20.8	28.5	39.7	19.5
11	9.6	14.1	18.6	21.1	41.8
12	4.4	6.1	6.8	6.2	9.9
13	11.2	10.7	9.0	6.1	5.5
14	12.1	9.0	8.3	4.2	0.8
15	34.8	30.6	15.9	4.4	1.6
16	2.6	1.7	1.6	1.7	3.5
17	33.0	46.1	63.9	20.4	12.6
18	25.6	36.1	25.2	27.2	32.3
19	2.4	3.1	5.1	10.3	39.7
20	0.9	0.7	0.1	0.0	0.8
21	1.0	1.7	2.2	2.0	2.1
22	8.0	7.0	6.8	2.5	4.1
23	28.6	32.0	19.8	16.0	2.3
24	18.9	20.9	22.9	31.7	36.3
25	16.9	26.6	44.0	62.2	53.1
26	6.7	8.7	13.1	20.0	30.8
27	42.8	36.8	29.4	24.6	24.9
28	24.4	27.7	6.9	2.2	0.5
29	2.4	5.4	9.5	18.1	28.5
30	9.9	8.7	7.9	9.1	13.4
31	50.4	50.1	47.5	49.3	46.1
32	8.6	9.0	11.0	14.8	26.8
33	2.1	1.8	1.8	4.2	17.8
34	7.6	15.4	26.6	28.9	40.0
35	8.4	14.0	16.5	11.3	10.5
36	0.5	0.4	0.1	0.6	0.4
37	30.7	33.2	17.2	14.3	7.3
38	18.4	17.9	17.1	15.0	21.4

areas. The number of significant areas decreases from 14 to 10 with an increase in the spatial averaging size from the area shown in Fig.7 to the largest area, referred to as Area + 4. This can be due to the basence of strong planetary-scale stationary waves in the Southern Hemisphere summer. The number of significant areas for predictability of averages for days 16-46 and 32-60 is too small to investigate its dependence on the size of the averaging area.

Table 2b. Values of   for mid-January and mid-February (days 16-46) for different sizes of the area of averaging

Region no.	(Area − 1)	Area (Fig.7)	(Area + 1)	(Area + 2)	(Area + 4)
1	4.3	3.9	5.5	5.5	4.9
2	4.3	5.9	5.5	5.3	2.9
3	5.9	4.6	2.4	2.3	2.2
4	2.2	1.7	1.5	0.8	0.5
5	1.5	1.6	1.4	0.8	0.7
6	1.7	1.2	1.0	0.8	0.6
7	0.2	0.2	0.3	0.2	0.1
8	29.3	28.3	24.7	16.5	7.9
9	5.9	5.5	4.9	4.6	3.0
10	1.2	1.0	0.5	0.0	0.4
11	1.6	1.8	1.5	1.7	2.9
12	1.2	1.8	2.9	4.2	4.4
13	0.2	0.1	0.3	0.9	4.6
14	1.4	2.4	1.8	1.5	0.9
15	4.2	4.5	2.5	1.3	2.2
16	3.0	3.8	3.9	3.1	1.9
17	0.2	0.0	0.1	1.4	1.2
18	0.7	0.6	0.4	1.0	3.4
19	3.3	5.7	7.8	11.5	16.3
20	3.1	3.2	2.8	2.8	0.3
21	2.0	2.0	2.8	3.2	0.2
22	1.8	1.3	1.0	0.1	1.2
23	11.3	12.8	2.0	1.2	0.4
24	4.4	9.0	15.8	19.1	7.7
25	3.0	3.4	3.6	3.6	2.6
26	3.5	4.0	5.5	7.2	6.7
27	27.4	25.0	22.1	19.6	10.8
28	10.7	11.0	0.8	0.1	0.3
29	1.8	2.1	4.2	5.6	4.3
30	0.5	0.1	0.0	0.0	0.2
31	0.5	0.7	1.0	1.2	1.9
32	0.1	0.1	0.4	0.7	1.8
33	4.0	3.8	3.6	4.0	4.6
34	4.9	9.6	16.1	16.3	17.2
35	1.8	2.5	3.4	6.9	2.9
36	0.0	0.2	0.5	0.7	1.1
37	7.9	8.0	10.0	8.4	4.5
38	9.1	11.0	11.1	9.0	7.2

From these results it can be conjectured that at the 95% confidence level there is almost no predictability for the averaging period of days 32-60; there is a substantial degree of predictability for days 1-31 but there is only partial predictability for days 16-46. This suggests that the observed initial conditions, together with their instabilities and nonlinear interactions, remember themselves, at least up to a month, to such an extent that the monthly means for different integrations are significantly different from the monthly means of randomly perturbed initial conditions.

Table 2c. Values of F for February (days 32-60) for different sizes of the area averaging

Region no.	(Area - 1)	Area (Fig.7)	(Area + 1)	(Area + 2)	(Area + 4)
1	5.2	6.3	7.9	9.3	10.6
2	1.8	2.0	2.4	3.0	5.0
3	1.4	1.3	2.2	2.4	2.2
4	0.8	0.9	1,3	2.7	12.0
5	0.7	0.4	0.5	1.4	4.0
6	3.7	4.2	5.3	6.0	6.3
7	0.3	0.1	0.1	0.1	0.1
8	1.7	1.2	0.8	0.5	1.0
9	0.3	0.2	0.4	0.4	1.5
10	2.1	2.4	2.9	2.0	1.3
11	3.2	3.2	3.3	3.4	3.4
12	0.8	1.0	1.2	1.6	1.5
13	1.3	1.2	1.1	0.8	0.5
14	0.5	0.1	0.1	0.3	0.6
15	2.1	1.1	0.8	0.9	0.7
16	2.3	2.9	2.9	2.7	2.0
17	1.3	1.1	0.8	0.8	0.6
18	4.4	2.0	2.4	1.8	1.1
19	0.5	1.5	2.9	4.7	8.1
20	7.6	16.1	15.7	5.0	0.6
21	1.5	1.6	2.7	3.1	5.8
22	1.0	0.8	0.7	6.2	2.2
23	0.8	1,0	0.7	0.5	0.3
24	0.5	0.7	1.2	1.3	0.9
25	4.6	5.1	3.6	2.5	1.1
26	2.2	3.0	4.0	3.9	2.1
27	7.5	7.3	7.3	7.2	5.1
28	4.1	4.3	1.3	0.4	0.1
29	0.8	0.6	1,0	1.1	1.0
30	0.3	0.4	0.4	0.3	0.3
31	0.0	0.0	0.1	0.1	0.3
32	0.6	0.6	0.5	0.4	0.2
33	4.0	3.2	2.0	1.5	0.6
34	1.6	0.8	0.3	0.4	1.5
35	2.3	1.6	1.6	0.2	0.4
36	0.0	0.1	0.2	0.1	3.3
37	1.1	0.8	1.5	1.6	1.1
38	3.7	5.0	5.3	5.3	3.1

We also have calculated the values of F at each grid point for five different 30-day means corresponding to days 1-30, 8-37, 15-44, 22-51 and 29-58. Figures 9a-9e give the plots of F for the 500 mb geopotential height over the Northern Hemisphere. The dotted areas are significant at the 95% level and the cross-hatched areas are significant at the 99.5% level. A systematic degradation of predictability is found to occur from the first to the fifth map. Most of the grid points are significant at the more than 95% confidence level for the predictability of the first 30-day mean. This is a

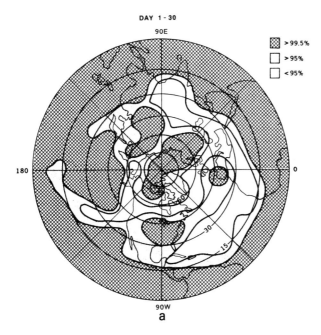

Fig.9a. F values for 500 mb geopotential height field, calculated at each grid point, for 30-day mean (days 1-30). Light-shaded areas are significant at 95%, and cross-hatched areas are significant at 99% confidence level. Significance level for blank level is <95%

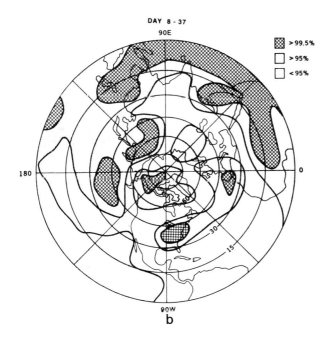

Fig.9b. As in Fig.9a except for the 30-day mean (days 8-37)

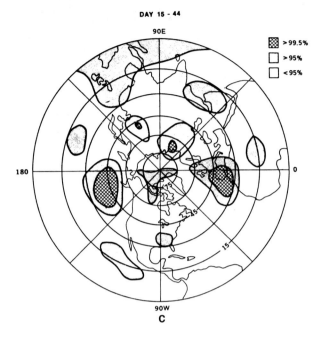

Fig.9c. As in Fig.9a except for the 30-day mean (days 15-44)

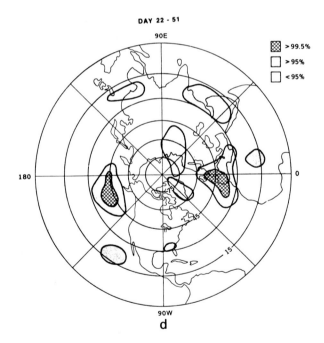

Fig.9d. As in Fig.9a except for the 30-day mean (days 22-51)

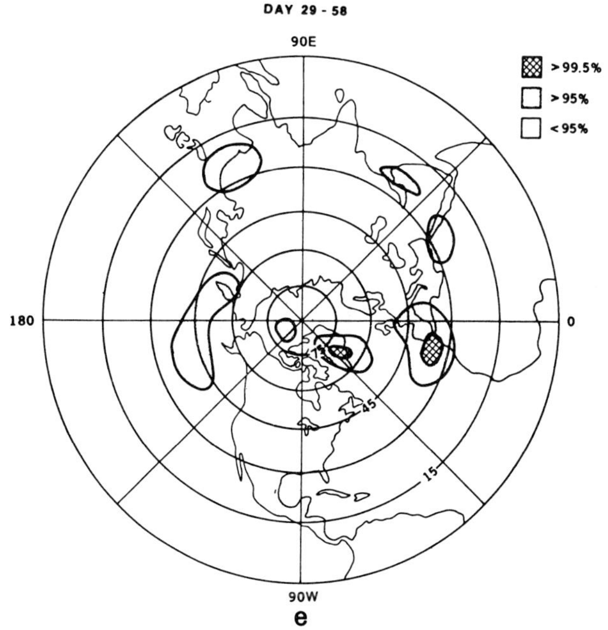

DAY 29-58

Fig.9e. As in Fig.9a except for the 30-day mean (days 29-58)

clear indication of the dominance of initial conditions which were different, and it also indicates that the initial planetary-wave configurations could not be drastically changed by the random perturbations. A more encouraging conclusion can be drawn from the distribution of F values for the 30-day mean of days 8-37 and 15-44. Even after ignoring the first 7 and 14 days of integration, which are considered to be the "useful" and "theoretical" upper

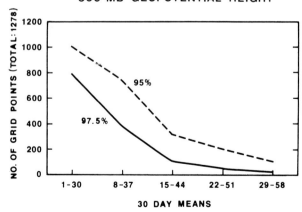

Fig.10. Number of grid points for which F value is significant at more than 95 and 97.5%

limits of synoptic-scale deterministic prediction, there are large areas for which the significance level is more than 95%. Figure 10 gives the number of grid points between 2 and 78°N which are significant at the 95 and 99% confidence levels for the five averaging periods. The number of grid points for the averaging period of days 29-58 is too small to be statistically significant and it may be reasonable to conclude that the model used in this study, *with climatological mean boundary conditions*, may not have any success in predicting the monthly mean of the second month.

6. Discussion of Results

Since the results of any such study are bound to be model dependent, it is necessary to examine especially those characteristics of the model which may have a bearing on the results and their interpretation. For example, if integrations were started with largely different initial conditions for a hypothetical defective model, in which initial conditions persist throughout the course of the integration, one can get a very large value of F. This may lead to a false conclusion about the existence of predictability. In this section we will show that the day-to-day fluctuations simulated by the model are realistic and, except near the poles, comparable to the observed day-to-day fluctuations in the atmosphere.

Similarly, the reduction in the values of F for the second month could be caused by the model property, or more appropriately the model deficiency, that the model simulations for the second month converged to the same state and therefore their variance was reduced. We will show that this is not the case for the present model. The reduction in F occurred mainly due to very large departures between the control run and the perturbation runs for the second month of integration and not due to the relaxation of all the simulations to the same model state.

a. *Day-to-day Fluctuations*

In order to show that the large values of F for January were not due to the persistence of different initial conditions, we have examined the day-to-day fluctuations of sea level pressure, 500 mb geopotential height, and temperature for the nine models runs.

In order to determine the model's ability to simulate the day-to-day fluctuations, we have also calculated, at each grid point, the standard deviation of daily values for nine model runs for January and February.

If P_{ijkt} denotes the daily value of any meteorological variable (*viz.*, sea level pressure or 500 mb geopotential height) at grid point i, j for model run k on day t, we calculate a measure of day-to-day variability σ_{ij} at grid point i, j as

$$(\sigma_{ij})^2 = \sum_k \sum_t (P_{ijkt} - \bar{P}_{ijk})^2 / (K^*T)$$

where

$$\bar{P}_{ijk} = T^{-1} \sum_t P_{ijkt}$$

and

$k = 1, 2, \ldots, K \ (K = 9)$

and

$t = 1, 2, \ldots, T$

(T = 31 for January and 29 for February).

For comparison, we have also calculated σ_{ij} for 15 years of observation (K = 15). The observational data set consisted of the NMC (National Meteorological Center) analyses which were received from NCAR (National Center for Atmospheric Research). Figure 11 shows the zonal averages of σ_{ij} for nine model runs and 15 months of observations for sea level pressure for January and February. Results for 500 mb geopotential height field (not shown) also were similar except that the differences near the poles were even larger. The model tends to underestimate the day-to-day variability near the winter pole. This is not unrelated to a major model deficiency that gives an unrealistically cold upper troposphere and strong zonal wind. In the tropics and the midlatitudes, the day-to-day variability of the model is found to be realistic and quite comparable to the day-to-day fluctuations observed in the real atmosphere.

We have not compared the day-to-day predictions with the corresponding observations because numerical weather prediction was not the main intent of this study. Moreover, we have not used the observed boundary conditions for the corresponding initial conditions. The primary aim here was to determine the time limit for which the random perturbations can make the space-time averages completely unpredictable.

b. *Does the Model Relax to the Same State for the Second Month?*

Since the external forcings (solar heating and climatological mean boundary conditions) for all the model runs are identical, it is very likely that all

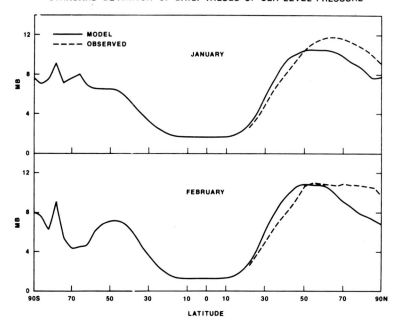

Fig.11. Zonally averaged standard deviation of daily grid point values for sea level pressure for January (upper panel) and February (lower panel), for model (solid line) and observations (dashed line)

the simulated model states will relax to the same mean climate. The relaxation time, in general, will be determined by the time scale of the external forcing and the nature of dynamical system which might have its own internal feedbacks which, in turn, may interact with the external forcing (*viz.*, cloud-radiation interaction). A comprehensive examination of this question is beyond the scope of the present study, especially because the projects of predictability of a few days to a month are largely, and sometimes solely, determined by the dynamical instabilities and their interactions. In the present study we investigate a limited aspect of the question. Does the reduction in the value of F for February occur due to similarity between all the model simulated states or is it due to very large differences between the control run and its perturbation run? In order to provide a qunatitative answer to this question, we have examined the following three aspects of the model simulations.

Table 3. Values of F_1 and F_2 for different regions for January and February

Region no.	F_1 January days 1-31	F_1 February days 32-60	F_2 January days 1-31	F_2 February days 32-60
1	41.7	31.8	12.1	14.9
2	11.6	16.4	10.0	23.9
3	8.9	8.2	1.4	13.5
4	11.3	2.1	1.1	6.6
5	0.4	0.2	5.1	1.5
6	1.8	10.2	5.8	7.2
7	14.6	0.6	8.5	10.1
8	58.8	4.9	7.9	11.6
9	9.2	0.7	1.0	7.4
10	0.7	0.9	0.1	1.1
11	0.3	1.0	0.0	0.9
12	0.5	0.8	0.2	2.3
13	0.3	0.2	0.1	0.6
14	1.1	0.0	0.3	1.7
15	2.5	0.2	0.2	0.7
16	0.1	0.7	0.3	0.7
17	1.0	0.5	0.0	1.6
18	1.3	0.4	0.1	0.7
19	1.0	0.2	0.9	0.5
20	0.2	1.4	1.1	0.2
21	1.2	1.3	2.1	2.4
22	1.3	1.7	0.5	5.9
23	77.8	9.0	7.2	26.3
24	6.7	1.7	0.9	7.2
25	26.8	3.8	3.0	2.2
26	41.5	13.3	14.3	13.3
27	19.9	26.3	1.6	10.7
28	195.8	47.5	21.1	32.6
29	21.4	10.5	11.9	50.5
30	30.8	3.7	10.6	25.0
31	85.6	0.3	5.1	64.1
32	103.0	34.6	34.3	174.3
33	5.7	27.4	9.2	25.7
34	79.1	4.3	15.3	14.8
35	26.0	13.7	5.5	24.5
36	1.7	8.7	12.3	138.7
37	125.0	29.8	11.2	111.4
38	226.7	122.7	37.0	72.4

1) Change in F_1 and F_2 From January to February

Table 3 gives the values of F_1 and F_2 for January and February. If all the model simulations relaxed to the same state for the second month, the value of F_2, which measures variability among the control and their perturbation runs, would be very small. However, it can be seen that, from January to February, the value of F_2 has increased for 32 out of 38 areas. Most of the reduction in F from January to February is due to increase in value of the denominator (F_2). This supports the contention that the loss of predictability for the second month is not due to relaxation of the model to a mean state but due to large differences in the evolution of the dynamical system represented by the model.

Table 3 also shows that the values of F_1 have decreased for 26 out of 38 areas (although the decrease for most of the areas is very small). As possible reason for the decrease in F_1 could be a model deficiency, as mentioned above, that all model states were similar. We have shown above that this is not the case because F_2 has increased. Since F_1 is a measure of variability among the different box means, large deviations of control and perturbation runs will lead to reduction in the value of F_1 from January to February. It is important to note that this indeed is the case for the present model. If all the model simulations for the second month relaxed to the same state, the prospects for dynamical predictability with this particular model would have been quite hopeless. However, since the lack of predictability for the second month is due to the large deviations between the dynamically evolving flow fields, it is conceivable that, with the use of more realistic dynamical models and accurate physical parameterizations, the limit of predictability of time averages may be extended beyond one month.

2) Variability Among the Control Runs for January and February

We know that the initial conditions for the three control runs were as different as three randomly chosen years and, therefore, standard deviation among them gave a measure of interannual variability. If these initial conditions relaxed to a very similar model state for the second month, the standard deviation for the second month (February) should be smaller than the interannual variability. Figure 12 shows the zonally averaged standard deviation among the three control runs for January and February for 500 mb geopotential height and sea-level pressure. It is seen that the variability among the three control runs for the second month is indistinguishable from the variability for the first month. This provides an additional support to the contention that the reduction in F and the loss of predictability of the second month is not due to the relaxation of the model to the same state, but due to large errors between the control and the perturbation runs.

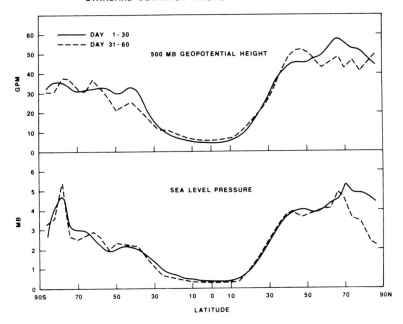

Fig.12. Zonally averaged standard deviation among monthly values for three control runs for first 30-day mean (solid line) and next 30-day mean (dashed line), for 500 mb geopotential height (upper panel) and sea level pressure (lower panel)

3) Error Between Control and Perturbation runs

We also have calculated the magnitude of the error between a control run and its perturbation runs. The differences between the control and its perturbation increase steadily with an increase in the length of integration, and we have shown that the magnitude of this difference for the second month is large enough to be comparable to the differences between the observed monthly means. The error $E(\phi)$, which is a measure of the rms error among the monthly means due to random perturbations, for any latitude ϕ, is defined as

$$\{E(\phi)\}^2 = \frac{1}{2\pi}\int_0^{2\pi} \frac{1}{10} \left\{ \begin{array}{l} \{(M_{11}-M_{12})^2+(M_{11}-M_{13})^2+(M_{12}-M_{13})^2+(M_{21}-M_{22})^2 \\ +(M_{21}-M_{23})^2+(M_{21}-M_{24})^2+(M_{22}-M_{23})^2 \\ +(M_{22}-M_{24})^2+(M_{23}-M_{24})^2+(M_{31}-M_{32})^2 \} d\lambda \end{array} \right. ,$$

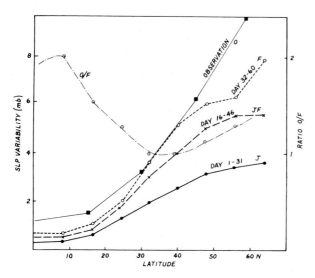

Fig.13. Zonally averaged rms error between control and perturbation runs averaged for days 1-31 (solid line), days 16-46 (long dashed line) and days 32-60 (short dashed line). Thin solid line is for the observed January sea level pressure for 16 years (1962-1977). Dashed-dotted line is the ratio of observed and model (days 32-60) variances

where M_{ji} refers to the grid-point value of monthly mean sea level pressure for run number C_{ji} (refer to Table 1) and integral over λ denotes integral over all longitudes (72 grid points for $d\lambda = 5°$) along the latitude ϕ.

Figure 13 shows the zonally averaged values of $E(\phi)$ for three different averaging periods. Dotted, dashed and solid lines, labeled as J, J/F and F refer to the monthly means for days 1-31, 16-46 and 32-60 respectively. The corresponding value of $E(\phi)$ among all possible combinations of 16 years (1962-1977) of observed monthly mean sea level pressure over the Northern Hemisphere are shown by a thin solid line. Except for the low latitudes, the average error among the control and the perturbations runs for days 32-60 is comparable to the observed differences between Januaries of different years. This clearly demonstrates that the model-simulated monthly means for the second month are not relaxing to the same value but they are perhaps as different as possible for the prescribed external forcing.

The dash-dot line in Fig.13 gives the ratio of observed $E(\phi)$ and the model $E(\phi)$ for February. In agreement with the results of Charney and Shukla (1980), it is found that the ratio is maximum in low latitudes. This supports their hypothesis that a large fraction of the interannual variability of the monthly means in the low latitudes may be related to the changes in the slowly varying boundary conditions.

c. *Model Limitations*

As mentioned earlier, a major deficiency of the present model is the underestimation of day-to-day variability near the poles. Variability of monthly means for days 32-60 also is smaller than the observed variability near the poles. Figure 14 shows the zonally averaged standard deviations for 16 years of observed January means, and monthly mean (days 32-60) for nine model runs. The largest discrepancy is found to occur near the winter pole. North of 60°N, the model variability is only ~60% of the observed variability. This is related to a major model deficiency that the upper tropospheric-lower stratospheric temperatures are very cold compared to the observations and the zonal winds at the upper levels are quite large. Removal of such systematic and serious deficiencies of the general circulation models will be a prerequisite for their utility as forecast models for dynamical prediction of monthly means.

It should be emphasized that we have subjected this model to a rather rigorous and perhaps even unfair test of its predictability by imposing climatological mean boundary conditions for integrations starting from observed initial conditions. Since the initial conditions are not quite consistent

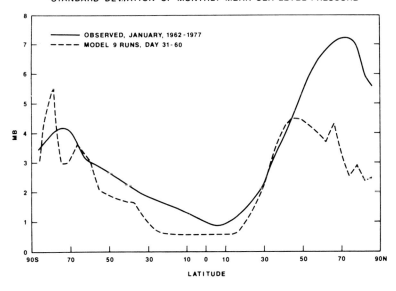

Fig.14. Zonally averaged standard deviation among monthly (days 31-50) sea level pressure for nine model runs (dashed line) and 16 observed Januaries (solid line)

with the imposed stationary forcings, and since the model does not simulate the stationary circulation realistically, spurious traveling wave components are generated and tend to make the model unpredictable from the very beginning of the integration. If all the integrations were started with the corresponding observed boundary conditions, it would reduce the amount of initial inconsistency between the initial dynamical state and the underlying forcing and therefore it might reduce the errors due to spurious traveling waves.

We also should summarize other known model deficiencies which may have a bearing on the present results and their interpretation. Straus and Shukla (1981) have shown that although the total variance simulated by the GLAS model is not too unrealistic, the cyclone wave variances are overestimated and the low-frequency planetary wave variances and the stationary variances are significantly underestimated. This will tend to reduce the variance among the monthly means. Since we have no reason to believe that the reduction in the variance among control runs will be smaller than the reduction in the variance among the perturbation runs, we conclude that this particular deficiency of the model may not alter our conclusions. On the other hand, since the cyclone-scale variances are overestimated in the GLAS model and since these scales are known to grow rapidly and contribute toward the unpredictability of larger scales, it will tend to overestimate the variance among the perturbation runs. Since the differences among the control runs is very large at the initial time itself, dynamic instabilities cannot make them any more different from each other. It is, therefore, reasonable to conclude that this particular deficiency of the GLAS model tends to underestimate the predictability, and if the model was more realistic in simulating the cyclone-scale variances, the theoretical upper limit of dynamical predictability would have been larger. It should be pointed out that one of the major motivations of carrying out this study, in contradistinction to a study in which model generated forecasts are compared with the observations, was to compare two different properties of the same model rather than comparing the model results, with the observations. Model results are generally compared with the observations to determine the skill of prediction and its operational usefulness. This is not the objective of this study. Here we are addressing a different question: Is there a basis to make dynamical prediction of monthly means?

Although we conclude from this study that monthly means are potential predictable with dynamical models, it remains to be seen whether the skill of such a prediction will be any better than that of a statistical prediction scheme. It is likely that a combined statistical-dynamical approach may have more skill than either a pure dynamical or a pure statistical approach.

7. Summary and Conclusions

The GLAS general circulation model is integrated for 60 days with nine different initial conditions. Three of the initial conditions are the observations on 1 January of 1975, 1976 and 1977. Six other initial conditions are obtained by adding random perturbations, comparable to the errors of observations, to the three observed initial conditions. For all the integrations, the boundary conditions of sea surface temperature, sea ice, snow and soil moisture are the same as their climatological values changing with season.

We have carried out an analysis of variance to compare the differences of monthly means predicted from largely different observed initial conditions to the monthly means predicted from randomly perturbed initial conditions. If the variability among the monthly means predicted from very different initial conditions becomes comparable to the variability for randomly perturbed initial conditions, it is concluded that the model does not have any predictability beyond that period.

It is found that the observed initial conditions in different years have large differences compared to the errors of observation, and different planetary-scale configurations retain their identity at least up to one month or beyond. The predictability limit for the synoptic-scale waves (wavenumbers 5-12) is only about two weeks, but for the planetary waves (wavenumbers 0-4) it is more than a month. We have shown that there is a physical basis to make dynamical prediction of monthly means at least up to one month.

For climatological mean boundary conditions, there is no potential for predictability in the second month (days 31-60) because the monthly means predicted from largely different initial conditions become indistinguishable from the monthly means predicted from random perturbation. It is encouraging to note, however, that the lack of predictability for the second month is not because all the model integrations converge to the same mean state but because of very large differences due to random perturbations in the initial conditions. This suggests that it is possible, at least in principle, to extend the predictability limit even beyond one month by improving the model, the initial conditions, and the parameterizations of physical processes.

In this paper we have examined only the dynamical predictability with climatological mean boundary conditions. There could be additional predictability due to fluctuations of the slowly varying boundary conditions of sea surface temperature, soil moisture, sea ice and snow. There is sufficient observational and GCM-experimental evidence to suggest that the fluctuations of sea surface temperature and soil moisture in low latitudes produce significant changes in the monthly and seasonal mean circulation. Likewise, if the anomaly of sea surface temperature, sea ice or snow is of large magnitude and of large-scale, it can produce significant changes in the mid-latitude circulation. Due to the presence of strong instabilities in the middle latitudes, boundary effects have to be sufficiently large to be significant. However, there seems to be some evidence that the fluctuations of the heat sources in the tropics can produce significant changes in the extratropical latitudes, and, therefore, there is potential for additional midlatitude predictability because of its interaction with low latitudes.

8. Suggestions and Further Considerations

One of the serious limitations of the present general circulation models seems to be their inability to simulate the stationary circulation. This problem seems to be common to dynamical weather prediction models also, because predicted fields show systematic geographically fixed error structures. The following factors may be cited as possible reasons for incorrect simulation of monthly mean fields:

1) *Inadequate horizontal resolution*: Errors in small scales may affect the large scales, and higher-order accuracy is needed for long waves.

2) *Inadequate vertical resolution, location of the upper boundary, and upper boundary condition*: Most of the models do not treat adequately the vertical propagation and damping of stationary waves.

3) *Inadequate treatment of orography*: Either due to coarse resolution or due to improper finite-difference treatment, flow over and around mountains is not well simulated.

4) *Diabatic heating processes over land and ocean*: Due to inadequate parameterization of boundary-layer, convection, cloud-radiation interaction and physical processes at ocean and land surfaces, the vertical structure

of the diabatic heating field is not realistic and therefore the forced stationary waves are not realistic. That, in turn, causes the amplitudes and locations of the storms and their tracks to be unrealistic. If the stationary component of the circulation is not simulated correctly, the discrepancy also will manifest itself into the propagating components of the circulation.

The results of the present study and the results of several other long-range prediction studies of Miyakoda (personal communication) suggest that the prospects of useful long-term dynamical prediction will largely depend upon our ability to develop highly accurate dynamical models with can realistically simulate the stationary and the transient components of the circulation.

The present generation of dynamical models already have demonstrated a degree of verisimilitude which justifies a systematic program to establish the feasibility of long-term dynamical prediction. What is needed first is a very systematic evaluation and intercomparison of dynamical models to determine the precise nature of their weaknesses in simulating the structure and amplitudes of stationary and transient circulations. This is beyond the scope of any individual research scientist and it requires an institutional and organizational framework to begin such a massive undertaking. It is perhaps fair to say that the physical basis and justification to initiate a program of dynamical long-range prediction today is at least as strong as was the basis to start numerical weather prediction 25 years ago, and likewise, a systematic program of long-range dynamical prediction will undoubtedly increase our understanding of the dynamics of the atmosphere and ocean.

Acknowledgments. I am grateful to Professors J.G. Charney and E.N. Lorenz for the benefit of many useful discussions and suggestions during the course of this study. I wish to express my deep appreciation for discussions with and comments by Drs. M. Halem, R. Hoffman, Y. Mintz, K. Miyakoda, D. Randall, D. Straus, D. Gutzler and Prof. M. Wallace. I am thankful to Dr. D. Straus for providing the data on the observed interannual variability of variances. It is a pleasure to express my appreciation to Tom Warlan and Ricky Sabatino for numerical calculations, Karen DeHenzel and Debbie Boyer for typing the manuscripts, and Laura Rumburg for drafting the figures.

References

Charney, J.G. (1960): Numerical prediction and the general circulation. *Dynamics of Climate*. R.L. Pfeffer (ed.). Pergamon, 12-17

Charney, J.G., Devore, J.G. (1979): Multiple flow equilibria in the atmosphere and blocking. J. Atmos. Sci. 36, 1205-1216

Charney, J.G., Shukla, J. (1980): Predictability of monsoons. *Monsoon Dynamics*. Sir James Lighthill and R.P. Pearce (eds.). Cambridge University Press

Charney, J.G., Straus, D.M. (1980): From-drag instability, multiple equilibria and propagating planetary waves in baroclinic, orographically forced, planetary wave system. J. Atmos. Sci. 37, 1157-1175

Charney, J.G., Fleagle, R.G., Lally, V.E., Riehl, H., Wark, D.Q. (1966): The feasibility of a global observation and analysis experiment. Bull. Amer. Meteor. Soc. 47, 200-220

Charney, J.G., Shukla, J., Mo, K.C. (1981): Comparison of a barotropic blocking theory with observation. J. Atmos. Sci. 38, 762-779

Halem, M., Shukla, J., Mintz, Y., Wu, M.L., Godbole, R., Herman, G., Sud, Y. (1980): Climate comparisons of a winter and summer numerical simulation with the GLAS general circulation model. GARP Publ. Series 22, 207-253

Hays, W.L. (1963): *Statistics*. Holt, Rinehart and Winston, 719 pp

Leith, C.E. (1973): The standard error of time-averaged estimates of climatic means. J. Appl. Meteor. 12, 1066-1069

Leith, C.E. (1975): The design of a statistical-dynamical climate model and statistical constraints on the predictability of climate. The Physical Basis of Climate and Climate Modeling, GARP Publ. Ser. No. 16, Appendix 2.2

Lorenz, E.N. (1965): A study of the predictability of a 28-variable atmospheric model. Tellus 17, 321-333

Smagorinsky, J. (1969): Problems and promises of deterministic extended range forecasting. Bull. Amer. Meteor. Soc. 50, 286-311

Strauss, D.M., Halem, M. (1981): A stochastic-dynamical approach to the study of the natural variability of the climate. Mon. Wea. Rev. 109, 407-421

Strauss, D.M., Shukla, J. (1981): Space-time spectral structure of a GLAS general circulation model and a comparison with observations. J. Atmos. Sci. 38, 902-917

Predictability of Time Averages:
Part II: The Influence of the Boundary Forcings

J. SHUKLA

Abstract

We have discussed the physical mechanisms through which changes in the boundary forcings of SST, soil moisture, albedo, sea ice, and snow influence the atmospheric circulation. The slowly changing boundary forcings can increase the predictability of monthly means because their effects on quasi-stationary flow patterns and statistics of synoptic scale disturbances appears to be potentially predictable. Changes in the boundary forcings produce changes in the moisture sources and diabatic heat sources which in turn change the atmospheric circulation. The magnitude and the structure of the atmospheric response due to changes in any boundary forcing depends upon the existence of a suitable large scale flow which can transform the boundary forcing into a three dimensional heat source, which in turn can change the large scale flow and its stability properties. The structure of the large scale flow also affects the propagation characteristics of the influence which determines whether the effect is local or away from the source.

We have presented results of numerical experiments conducted with the GLAS climate model to determine the sensitivity of the model atmosphere to changes in boundary conditions of SST, soil moisture, and albedo over limited regions. It is found that changes in SST and soil moisture in the tropics produce large changes in the atmospheric circulation and rainfall over the tropics as well as over mid-latitudes. Although the area occupied by the land surfaces is small compared to the ocean surfaces, the fluctuations of soil moisture can be very important because the diabatic heat sources have their maxima over the land, and therefore, even small fluctuations of soil moisture can produce large changes in the total diabatic heating field. The natural variability due to day to day weather fluctuations is very large in the middle latitudes, and therefore, changes in the mid-latitude atmospheric circulation due to changes in the boundary forcings at middle and high latitudes have to be quite large to be significant. It is suggested that large scale persistent anomalies of SST, snow and sea ice, under favorable conditions of large scale flow, can produce significant changes in the mid-latitude atmospheric circulation. It is also likely that time averaged mid-latitude circulation can have additional predictability due to the influence of tropical boundary forcings.

We have also presented observational evidence to show that interannual variability of atmospheric fluctuations is significantly different from the intra-annual variability, and therefore, we conclude that part of the inter-annual variability is due to the influence of boundary forcings. Since the tropical spectra is found to be redder compared to the mid-latitude spectra, the tropical flows may be potentially more predictable.

1. INTRODUCTION

In a previous paper (Shukla, 1981) we examined the predictability of the initial conditions without external forcings, and it was shown that the dynamical predictability of the observed planetary wave configurations is sufficiently long that the predicted monthly means are significantly different from the monthly means due to random perturbations in the initial conditions. It was further suggested that in addition to the dynamical predictability of the monthly means there can be additional predictability due to the influence of the boundary forcings. This paper examines the influence of slowly varying boundary forcings at the earth's surface in determining the monthly mean circulation anomalies in the atmosphere. Figures 1a and 1b show the schematic

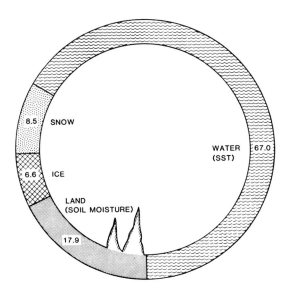

JANUARY

Figure 1a. Schematic representation of the atmosphere's lower boundary. Numbers denote the percentage of earth surface area covered by different boundary forcings during January

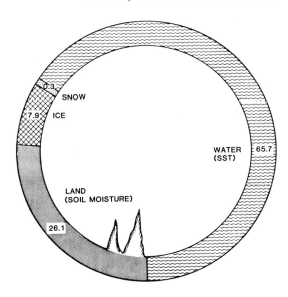

JULY

Figure 1b. Schematic representation of the atmosphere's lower boundary. Numbers denote the percentage of earth surface area covered by different boundary forcings during July

distribution of sea surface temperature (SST), soil moisture, surface albedo, snow cover and sea ice. There is sufficient observational evidence to assume that the rate of change of anomalies of these boundary forcings is slower than the corresponding atmospheric anomalies, and therefore for a limited period of time (viz. a month or sometimes even a season) these can be considered as "external" forcing to the atmospheric circulation. In the present paper, we will confine our discussion only to these five boundary forcings. A discussion of the influence of truly external forcing due to solar variability is beyond the scope of the present paper. Similarly, effects of slow changes in the composition (viz. CO_2) and optical properties of the atmosphere will not be considered because they appear to be too slow to affect monthly prediction.

One straightforward approach to determining the influence of boundary conditions on monthly mean prediction could be to carry out actual forecast experiments using the observed initial and boundary conditions. To our knowledge, no such forecast experiment has yet been carried out in which the <u>global distribution</u> of observed boundary forcings was used. It is only recently that global observing systems capable of determining global boundary conditions have been established, and still it is a formidable (but feasible) task to determine a global distribution of accurate boundary forcings. Another approach to determine the influence of the boundary forcings is to carry out a series of controlled sensitivity experiments with realistic general circulation models and to determine the regional distributions of time averaged response. This has been done using a variety of models and a variety of anomalous boundary forcings. A review of all such experiments carried out by many investigators is not possible in this paper. Here we summarize the results of only a few selected numerical experiments carried out with the Goddard Laboratory for Atmospheric Sciences (GLAS) climate model.

It should be pointed out that most of the numerical experiments carried out in the past were not especially designed to investigate the question of monthly predictability and, therefore, there was no detailed examination of the response during the first 30 days. The main interest was the difference in the simulated mean climate after the model has equilibrated to the altered boundary forcings. Such experiments are also quite useful for understanding the basic mechanisms and establish the role of boundary forcings in the interannual variability of monthly and seasonal means. In earlier studies, the choice of the geographical location of the anomaly (of SST or sea ice, etc.) was mostly determined by an analysis of the past observations, and naturally the past observations were analysed only for those areas for which data were conveniently available. On the basis of such observational studies and subsequent numerical experiments, several preliminary conclusions have been drawn about the possible relationships between the boundary forcing anomalies and the atmospheric circulation. These studies do not rule out the possibility that SST anomalies over other areas (hereto unexamined) will not be equally important. In fact, it is quite likely that all the boundary forcing, if considered globally, will produce more realistic and systematic response in the atmospheric circulation. One of the main contentions of this paper is to emphasize that further sensitivity and predictability studies using observations, simple linear models and complex GCMs should also be carried out for global distributions of boundary forcing anomalies.

Before presenting the results of actual numerical experiments, we will briefly mention some mechanistic considerations which suggest a physical basis for the influence of the boundary forcings.

1) Changes in the boundary forcings directly influence the location and intensity of the diabatic heat sources which drive the atmospheric circulation. The forcing at the boundary itself is generally not sufficient to produce significant changes in the atmospheric circulation; however, under favorable conditions of large scale convergence and divergence, the boundary effects get transmitted to the interior of the atmosphere and a thermal boundary forcing

gets transformed into a three-dimensional heat source which can be quite effective in influencing the dynamical circulation. The effectiveness of a boundary forcing in changing the atmospheric circulation therefore strongly depends upon its ability to produce a deep heat source and the ability of this influence to propagate away from the source. Since both of these factors are determined by the structure of the large scale dynamical circulation itself, the response of a given boundary forcing can be very different depending upon its size and geographical location, and upon the structure of the large scale circulation.

2) Boundary forcings of SST, soil moisture, surface albedo, snow and ice not only affect the heat sources and sinks, but they also affect the sources and sinks of moisture, which in turn affect the latent heat sources.

3) Existence of nonlinear multiple equilibrium states for a prescribed external forcing suggests that even weaker anomalies of boundary forcings, under favorable conditions, can produce significant anomalies in atmospheric circulation, and therefore actual response may be stronger compared to the one estimated from linear theories.

4) It is known that the inconsistency between the observed initial conditions and the prescribed stationary forcings due to mountains and diabatic heat sources can manifest itself as erroneous propagating transient components (Lambert and Merilees, 1978; Shukla and Lindzen, 1981). It is therefore desirable that for actual dynamical prediction from observed initial conditions, the observed global distribution of boundary forcings be used correctly. This will reduce the inconsistency between the initial conditions and the forcing, and therefore can reduce the growth of error of prediction.

In Part I we examined predictability for prescribed nonfluctuating boundary forcings, and the predictability was determined by error growth rate and error saturation value. We can identify the following reasons due to which changes in boundary forcings can influence these classical predictability parameters.

a) Changes in the boundary forcings can change the intensity and geographical location of synoptic scale instabilities. For example, changes in the amplitudes and phases of planetary waves can affect the storm tracks, which can affect the error growth and predictability in a particular region.

b) Changes in the boundary forcings can alter the saturation value of the error. As described earlier, the saturation value of the error depends upon the equilibration mechanisms which are different for different circulation regimes and also for different scales of instabilities. Boundary forcings can alter the equilibration level of the dominant fluctuations, resulting in a significant change in the time averaged mean circulation. It should be pointed out that the effects can be very different for different parameters. For example, if for a given value of boundary forcing, the amplitude of the wave disturbances is much less than its amplitude for another value of the boundary forcing, and if heavy rain or snow falls only during half of the life of each wave disturbance, and if the rate of rainfall is proportional to the intensity of the disturbance, then although the time averaged value of pressure or temperature will not be very different for two values of boundary forcing, the rainfall will be very different. This suggests that sometimes it may be quite useful to predict only the variance of a particular parameter.

2. SENSITIVITY OF MODEL ATMOSPHERE TO CHANGES IN BOUNDARY FORCINGS

In this section, we have summarized the results of several sensitivity studies carried out with global general circulation models to determine the influence of prescribed changes in SST, soil moisture, surface albedo, snow and sea ice, etc.

2.1 Sea Surface Temperature

The factors which determine the influence of sea surface temperature (SST) anomalies can be briefly summarized as follows.

(i) <u>The magnitude and the spatial and temporal structure of the anomaly</u>. Considering the linear response of the atmospheric system to diabatic forcing, the magnitude and the spatial scale of the anomaly can directly affect the response to the atmosphere. Large, persistent anomalies can produce a larger response than small fluctuating anomalies. The magnitude of the anomaly is also important in determining the nonlinear increase of evaporation and sensible heat fluxes.

(ii) <u>Normal sea surface temperature.</u> Due to the nonlinearity of the Clausius-Clapeyron equation, a 1° positive anomaly over a normal SST of 30° produces a much larger change in saturation vapor pressure than the same 1° anomaly superimposed upon a normal temperature of 20°. It is partly for this reason that anomalies in low latitudes can produce larger responses compared to similar anomalies in middle latitudes. The final response also depends upon the background SST field.

(iii) <u>The latitude of the anomaly</u>. Because of the smallness of the Coriolis parameter, horizontal temperature gradients in the tropics produce a much larger thermal wind than in middle latitudes. Due to the lack of geostrophic constraint in low latitudes, thermal anomalies produce much larger convergence than in middle latitudes (Hoskins and Karoly, 1981; Webster, 1981).

(iv) <u>Circulation regime</u>. The potential response of a given anomaly strongly depends upon the structure and dynamics of the circulation regime in which the anomaly is embedded. For example, a warm anomaly in the areas of large scale convergence (viz. ascending branches of Hadley and Walker cells) will be more effective than a comparable anomaly in the area of divergence. Similarly, in the middle latitudes the effect of a SST anomaly will strongly depend upon the location of the anomaly with respect to the phase of the prevailing planetary wave configurations.

(v) <u>Instability mechanism.</u> The time required for the atmosphere to feel the effect of the SST anomaly also depends upon the most dominant instability mechanism which determines the generation of a deep heat source due to surface anomaly. In tropical latitudes where CISK is the primary driving mechanism, a conditionally unstable atmosphere may respond rather quickly to a warm SST anomaly, whereas in mid-latitudes where the primary driving mechanism is baroclinic instability a given SST anomaly would affect the vertical shear and, therefore, the growth rates of baroclinically unstable waves.

(vi) <u>Structure of zonal flow</u>. Once a SST anomaly has produced a heat source, the structure of the prevailing zonal flow is of crucial importance in determining the propagation characteristics of the disturbances produced by the heat source. Tropical influences can affect the mid-latitude circulation by Rossby wave propagation or by changing the intensity of the Hadley cell and mid-latitude zonal flows which can interact with the mid-latitude thermal and orographic forcings.

In summary, the influence of boundary anomalies on the atmospheric circulation depends upon the existence of a favorable dynamical environment in which the surface forcing can be transformed into a three-dimensional heat source and the ability of this influence to propogate away from heat source. A warm anomaly in the tropics enhances evaporation and increases the moisture flux convergence which is the main contributor to the enhanced precipitation over the anomaly. Increased evaporation lowers the lifting condensation level, increases the buoyancy of the moist air, accelerates the deep convective activity, and increases the latent heating of the atmospheric column which further reduces the surface pressure and enhances moisture convergence. In the tropics, SST anomalies can also produce considerable effects away from the anomaly by modifying the areas of convergence and divergence. For example, if the intertropical convergence zone remains stationary over a very warm SST anomaly for a considerable length of time, those areas where ITCZ would have moved in its normal seasonal march will experience severe droughts. Similarly, a warm SST anomaly can change the location and intensity of the Walker circulation and enhanced ascending motion associated with the warm SST anomaly can produce reduced precipitation in the adjoining areas. This suggests that the effects of SST anomalies can be very nonlinear for particular regions, depending upon the location of the region with respect to the ascending branches of Hadley and Walker circulations. This nonlinearity can disappear for averages over very large areas. SST anomalies can also affect precipitation over distant areas by altering the moisture supply for the region.

The influence of SST anomalies for mid-latitudes is different than that for the tropics. SST anomalies in mid-latitudes, if they can produce a deep heat source, can change the quasi-stationary wave patterns which in turn can affect the location and intensity of the storm tracks. Therefore, the mid-latitude SST anomalies have a considerable potential to produce distant effects. This potential, however, is not fully realized for several reasons. The normal ocean temperature in mid-latitudes is relatively cold; the moist convection is not well organized and efficient enough to produce deep heat sources; due to strong geostrophic balance even large gradients in temperature do not produce large convergence, and finally, the natural variability of the mid-latitude atmosphere is so large that small effects due to boundary forcings cannot be distinguished from synoptic weather fluctuations.

Most of the observational and numerical studies that have been carried out so far have only looked into the influence of regional SST anomalies. We do not have enough observational and numerical experimental results to describe the effects of hemispheric and global scale SST anomalies. We hope that future studies will examine the effects of very large scale SST anomalies. Here we shall present the results of a few sensitivity studies carried out by global general circulation models to determine the influence of regional SST anomalies.

2.1.1. Effect of Arabian Sea sea surface temperature anomaly on Indian monsoon rainfall

Several observational studies have suggested that SST anomalies over the Arabian Sea can be one of the important boundary forcings which determine the monsoon rainfall and therefore affects its interannual variability. A numerical experiment was carried out by Shukla (1975) using the Geophysical Fluid Dynamics Laboratory (GFDL) model to test the validity of observed correlations. A cold SST anomaly shown in Figure 2a (anomaly run) was imposed over the climatological

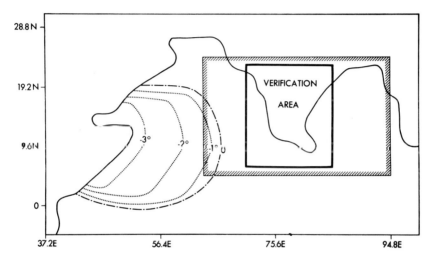

Figure 2a. Sea surface temperature anomaly (°C) over the Arabian Sea used for the GCM experiment and the area of verification

SST (control run) and the model was integrated for both cases. Figure 2b shows the model simulated rainfall of the Indian region for anomaly and control integrations. It is found that due to cold SST anomalies over the Arabian Sea monsoon rainfall is reduced over India. A similar experiment was carried out by Washington et al., (1977) using the National Center for Atmospheric Research (NCAR) model; however, their results were contrary to both the numerical results of the GFDL model and observed correlations. We have examined the results of

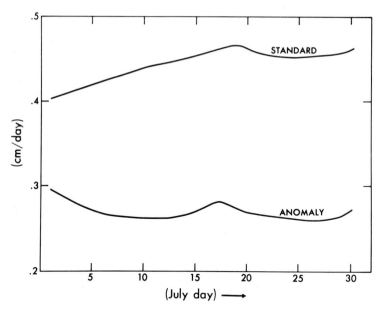

Figure 2b. 15 day running mean of the rate of precipitation (cm day^{-1}) averaged over the verification area for the standard and anomaly runs

the Goddard Laboratory for Atmospheric Sciences (GLAS) climate model in which a warm SST anomaly very similar in pattern to the anomaly shown in Figure 2a was used. Figures 3a and 3b show the model simulated rainfall for anomaly and control runs averaged over two areas shown in Figure 3c. The GLAS model was integrated for three different initial conditions, but very similar SST fields over the Arabian Sea. The curves labeled ANOMALY in Figures 3a and 3b show the average rainfall for three anomaly runs and the vertical bars denote the standard deviation among the three runs. It can be seen that warm SST anomalies over the Arabian Sea, as prescribed in these experiments, can produce enhanced monsoon rainfall over India. We do not know, however, whether such large anomalies are actually observed.

The apparent disagreement between the results of Washington et al., (1977) using the NCAR model, and our results using the GFDL and GLAS models can be explained by examining the low level monsoon flow as simulated by the three models and shown in Figure 3d. The low level monsoon flow as simulated by GFDL

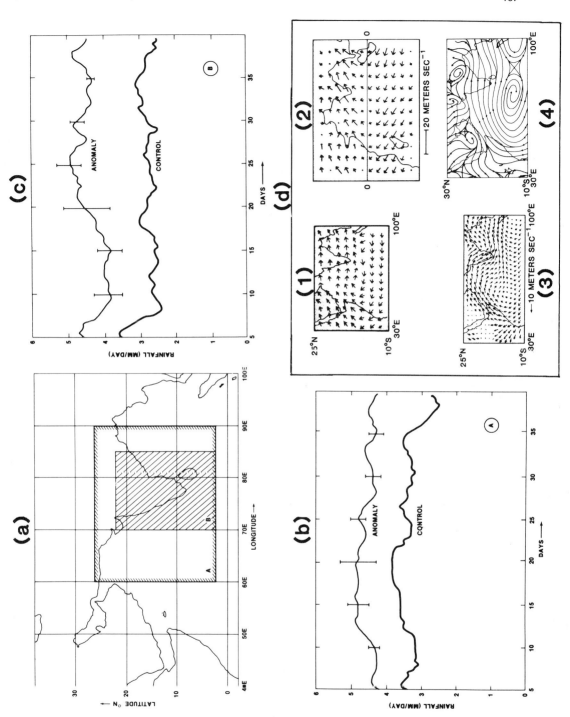

Figure 3. Panels (b) and (c) show the daily values of rainfall for control and anomaly runs averaged over areas A and B shown in panel (a). Length of the error bars represent standard deviation among three anomaly runs. Panel (d) shows the low level wind field for: (1) GLAS model, (2) GFDL model, (3) NCAR model, and (4) Observations

and GLAS models are more realistic compared to the NCAR model. In the NCAR model simulation, the air parcels flowing over the Arabian Sea hardly reach the Indian region as there is a strong but unrealistic southward flow before the monsoon current reaches the Indian coast. This could be one of the possible reasons why the NCAR model did not show significant response over India. This example illustrates the importance of a realistic simulation of the mean climate by the dynamical model used for the sensitivity studies. In order to be able to detect the effects of changes in the boundary forcings, and for the possible use of such models for prediction of time averages, the model should be able to simulate the mean climate accurately.

2.1.2. Effect of tropical Atlantic sea surface anomalies on drought over northeast Brazil

Moura and Shukla (1981) examined the monthly mean SST anomalies over the tropical Atlantic during March and rainfall anomalies over northeast Brazil during March, April and May. They found that the most severe drought events were associated with the simultaneous occurrences of warm SST anomalies over the north tropical Atlantic and cold SST anomalies over the south tropical Atlantic. They also carried out numerical experiments to test the sensitivity of the GLAS climate model to prescribed SST anomalies over the tropical Atlantic. It was found that the SST anomaly patterns, which resemble the observed ones during drought years, produced an intensified convergence zone (ITCZ), enhanced rainfall and low level cyclonic circulation to the north, and reduced rainfall and anticyclonic circulation to the south.

Figure 4a shows the SST anomaly used for the numerical experiment; the magnitude was chosen to be comparable to the maximum values observed during the 25 year period (1948-72). Figure 4b shows the 15 day running mean time series of daily rainfall averaged over the areas A and B (shown in Figure 4a) for control and anomaly runs. Area A includes the region of warm SST anomaly and area B contains the northeast Brazil region and the neighboring oceans with

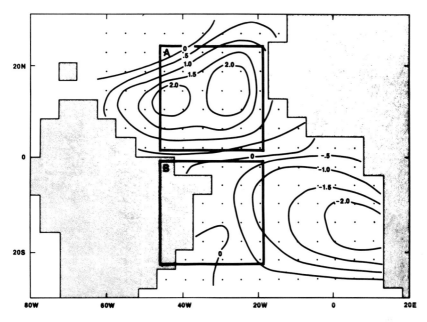

Figure 4a. Sea surface temperature anomaly (°C) over the tropical Atlantic used for the GCM experiment

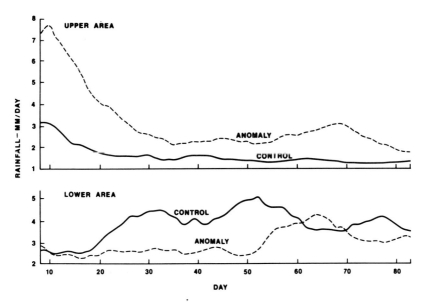

Figure 4b. 15 day running mean of daily rainfall (mm/day) for upper area (A) and lower area (B). (From Moura and Shukla, 1981)

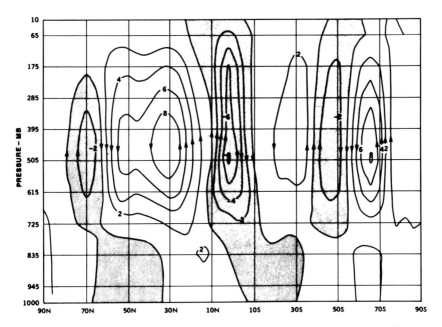

Figure 4c. Difference of 60 day mean meridional circulation (10^{13}gm/sec) averaged between the longitudes 50°W and 5°W. (From Moura and Shukla, 1981)

the cold SST anomaly. The rainfall over area A increases due to the warm SST anomaly and a shift of the ITCZ occurs from area B in the control run to area A in the anomaly run. Although the difference in rainfall is not systematic after 60 days, for days 20-60 the anomaly run has consistently less rainfall than the control run over northeast Brazil region. Figure 4c shows the difference of the first 60 day mean meridional circulation averaged between 50°W and 5°E. The anomalous meridional circulation shows an ascending branch with maximum vertical motions between 5° and 10°N and a descending branch to the south of the equator. A thermally direct local circulation is established with its ascending branch at about 10°N and its descending branch over northeast Brazil and adjoining oceanic regions. The driving for the anomalous circulation is provided by convection and latent heating associated with warmer SST anomalies over the northern tropical Atlantic, and cooling associated with

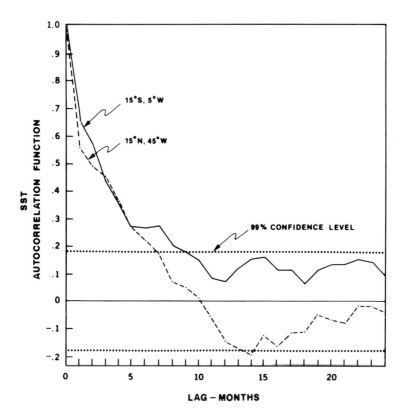

Figure 4d. Lag correlation function for monthly mean SST anomaly at 15°S, 5°W and 15°N, 45°W. (From Moura and Shukla, 1981)

colder SST anomalies in the southern tropical Atlantic. The combined effects of thermally forced subsidence and the reduced evaporation and moisture flux convergence produces severe drought conditions over northeast Brazil. It should be pointed out that descending motion influences a larger region to the east and west of northeast Brazil, however, the effect is most seriously felt over northeast Brazil. It was also noticed that the anomalous meridional circulations exhibited significant changes in the middle latitudes of both hemisphere.

Figure 4d shows the autocorrelation function for SST anomalies at 15°S, 5°W and at 15°N, 45°W. The strong persistence of SST anomalies in these two areas suggests the potential for prediction of drought over northeast Brazil.

2.1.3 Effect of Equatorial Pacific SST Anomaly on Tropical and Extratropical Circulations.

Horel and Wallace (1981) and Rasmusson and Carpenter (1982) have presented observational evidences of remarkable relationships between SST anomalies in tropical Pacific and a variety of atmospheric fluctuations including the Southern Oscillation and the Northern Hemispheric middle latitude circulations. "Warm episodes of equatorial Pacific SST anomalies are associated with the negative phase of the Southern Oscillation, weakening of easterlies in equatorial central Pacific, enhanced precipitation at equatorial stations east of 160°E, an intensified Hadley cell in Pacific sector, and a deepening and southward displacement of the Aleutian low" (Horel and Wallace, 1981). Analyses by Rasmusson and Carpenter (1982) have shown that during the month of December, a small warm SST anomaly appears along the Peru coast which rapidly increases to its peak value along the Peru coast during March and April of the following year. In the month of February of the following year, the warm SST anomaly disappears rapidly along the Peru coast, but the warmest SST anomalies are observed in equatorial central Pacific. It is these warm SST anomalies over the central Pacific, which occur over a climatologically warmer ocean surface which is also an area of large scale convergence, that produce a deep tropical heat source whose effects can propagate to the extratropical latitudes. Therefore, although the signal for a warm SST anomaly during February of a given year could be traced back to warm SST anomalies over the Peru coast fourteen months earlier, it is the enhanced heating associated with the SST anomaly during February that produces significant changes in the winter circulation of the northern mid-latitudes. Hoskins and Karoly (1981) used a multi-level linear primitive equation model to show that diabatic heat sources in the tropical regions can produce significant stationary responses in the middle latitudes if the zonal flow is favorable for the propagation of Rossby waves.

Shukla and Wallace (1983) have conducted sensitivity experiments with the
GLAS climate model to study the response of SST anomalies in equatorial Pacific.
Figure 5a shows the average of SST anomalies observed during the months of
November, December and January of 1957-58, 1965-66, 1969-70, and 1972-73,
provided to us by Dr. Rasmusson of NOAA. The GLAS climate model was integrated
with (anomaly run) and without (control run) the SST anomalies shown in Figure 5a.

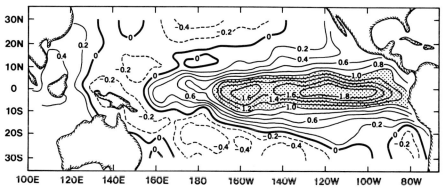

Figure 5a. Sea surface temperature anomaly (°C) used for the GCM experiment

Figure 5b shows the difference (anomaly minus control) field for 300 mb geopotential height averaged for days 11-25. It is seen that a difference of about 300 meters over North America and about -90 to -150 meters over Pacific and Atlantic is in agreement with results of observational studies as well as the results of linear models. Upper level highs to the north of the anomaly and a series of lows and highs further north and west are manifestations of Rossby waves propagating from the heat source. An examination of day-to-day changes from day 1 through 30 showed that this particular pattern was already established during days 5-10. However, it takes several days before model physics can generate a deep heat source above the warm SST anomaly. It should be pointed out that although the results of the linear models and this general circulation model are similar, there is a basic difference for their applicability to

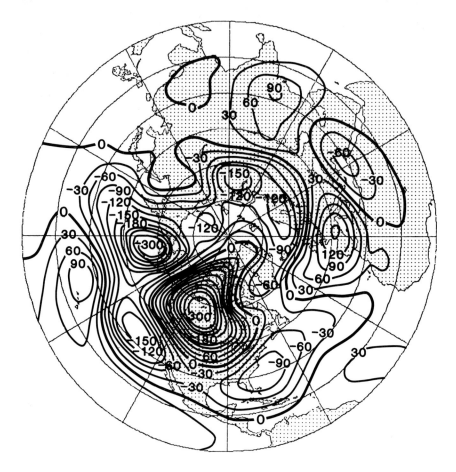

Figure 5b. Difference (anomaly-control) of 300 mb geopotential height field averaged for days 11-25

dynamical prediction of monthly means. The simple models prescribe the diabatic heating fields, whereas the general circulation model generates a diabatic heating field due to the presence of the warm SST anomaly, and therefore for a model to be useful as a prediction tool, its physical parameterizations must be able to transform the surface boundary forcing into a deep heat source. As mentioned earlier, the propagation of the tropical influences depends upon the structure of the zonal flow (influences cannot propagate across the zero wind line). In the present experiment, although the SST anomaly was centered right over the equator, the zonal flow was favorable for the tropical effects to propagate to middle latitides.

We have also examined the changes in the model simulated Hadley and Walker cells due to the prescribed SST anomalies. Figure 5c shows the difference between the Hadley cell for the anomaly and the control run averaged for days 6-25. It is seen that the zonally averaged Hadley cell intensifies due to warm SST anomalies over the equatorial Pacific. The Hadley cell over the Pacific sector alone is intensified even more. It was also noticed that in association with stronger Hadley cells, the westerly zonal flow was also stronger between 20° and 30°N. Figure 5d shows the difference (anomaly minus control) for the model simulated Walker cells averaged between 6°N and 6°S for days 6-25. Anomalous ascending motion occurs between the longitude sector 160°E - 160°W

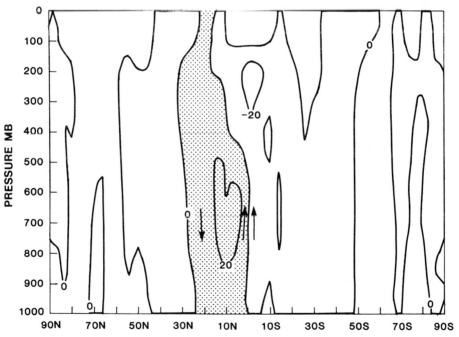

Figure 5c. Difference (anomaly-control) of Hadley cell (in units of 10^9 kg/sec) averaged for days 6-25.

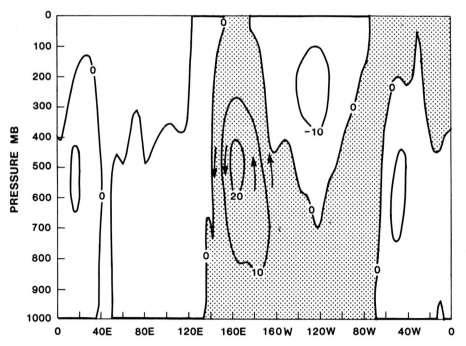

Figure 5d. Difference (anomaly-control) of Walker cell (in units of 10^9 kg/sec) averaged between 6°N-6°S and averaged for days 6-25

and descending motion occurs between the longitudes 140°E - 160°E. This is consistent with the observational evidence of enhanced precipitation east of 160°E being associated with warm SST anomalies.

A remarkable aspect of the results of this experiment is that the prescribed SST anomaly was an algebraic mean of four different episodes and therefore the prescribed maximum warm anomaly at any grid point was perhaps less than observed during any individual year. However, it produced significant changes within the first 30 days. The results also suggest that the SST anomalies can be of importance even for medium range (5-15 days) forecasting.

2.1.4 Effect of north Pacific SST anomaly on the circulation over North America.

As it was pointed out earlier, due to the large day-to-day variability in the middle latitudes, it is difficult to detect the influence of SST anomalies.

However, if a large scale SST anomaly of large magnitude persists for a long time, it can also produce significant effects in the midlatitude circulation. During the fall and winter of 1976-77, SST in the north Pacific was characterized by abnormally cold temperatures in the central and western portions of the northern Pacific with a warm pool located off the west coast of the U.S. (Figure 6a). Namias (1978) has suggested that the northern Pacific SST anomalies may have been one of the multiple causes of the abnormally cold temperatures in the eastern North America during the 1976-77 winter. We have carried out numerical experiments with the GLAS climate model to test this hypothesis. It should be noted that although the pattern of the SST anomalies observed during January 1977 was similar to the one shown in Figure 6a, the magnitude of the anomaly was only slightly more than half of the magnitude shown in the figure. It

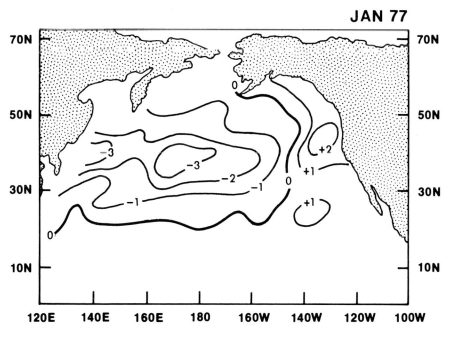

Figure 6a. Sea surface temperature anomaly (°C) over the north Pacific used for the GCM experiment. This anomaly pattern will be identical to the observed SST anomaly during January 1977 if the units are changed from (°C) to (°F)

should also be pointed out that during January 1977 a warm SST anomaly over the equatorial Pacific was also observed and it is likely that the circulation of North America could have been affected by combined effects of both anomalies. In this experiment, we have attempted to determine the effect of north Pacific SST anomaly only. We have integrated the GLAS climate model with six different initial conditions and climatological mean SST (control run) and for two of these initial conditions we have also integrated the model with imposed SST anomalies (anomaly run). The results that we show here are the differences between an average of six control runs and two anomaly runs.

Figure 6b shows the difference (anomaly minus control) for the 700 mb temperature averaged for days 15-45. As expected, negative temperature anomalies are found over the maximum negative SST anomaly and a warm 700 mb temperature anomaly is found over the warm SST anomaly. The most remarkable feature of this map, however, is the occurrence of two other centers of negative 700 mb temperature anomaly, one along 75°W and the other along 70°E. The cold 700 mb temperature anomaly over northeast North America is of largest magnitude. This feature suggests a planetary wave response due to north Pacific SST anomaly. The ratio between the temperature differences shown in Figure 6b and the standard deviation among the six control runs with climatological SST shows (Figure 6c) that the colder temperatures at 700 mb over northeast North America are as significant as the ones over the cold SST anomaly. The difference map (anomaly minus control) for the 500 mb geopotential height field shows (Figure 6d) that colder temperatures over northeast North America were due to anomalous northerly flow of cold air which in turn was caused by anomalous positive geopotential height anomalies to the west and negative geopotential height anomalies to the east. There are discrepancies between the simulated and the observed anomaly during the winter of 1977, because the response is one quarter wavelength out of phase. However, a detailed examination of the day-to-day evolution of the flow for the control and anomaly runs revealed that a persistent

TEMPERATURE AT 700 MB (DIFF)

Figure 6b. Temperature difference (°C) at 700 mb between the average of the two anomaly runs and the average of the six control runs, averaged for days 15-45

TEMPERATURE AT 700 MB (DIFF/SIGMA)

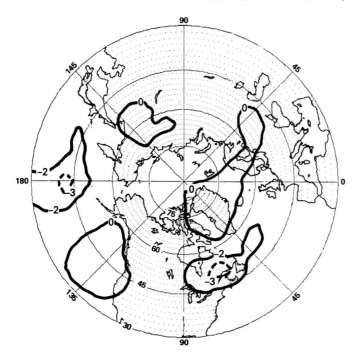

Figure 6c. Ratio of the temperature difference at 700 mb (shown in Figure 6b) to the standard deviation among 15-45 day averages of six control runs

GEOP HEIGHT AT 500 MB (DIFF)

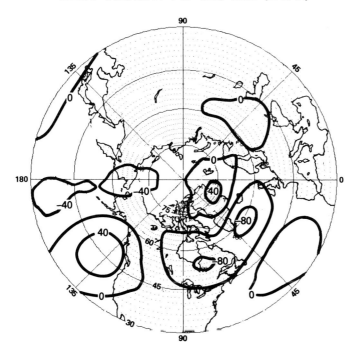

Figure 6d. Geopotential height difference (meters) at 500 mb between the average of two anomaly runs and the average of the six control runs averaged for days 15-45

blocking type of flow configuration existed in the anomaly run for more than 15 days (Chen and Shukla, 1983). Such persistent blocks did not occur in the control case. We are not quite sure if the blocking event generated in the anomaly run was due to presence of the SST anomaly. However, on the basis of the results of these experiments and the results of several observational studies, it is our conclusion that for favorable structures of the large scale flow, large scale SST anomalies in the mid-latitudes can be important in determining the anomalies of the mid-latitude circulation.

2.2 Soil Moisture

The annual average rainfall for the global continents is estimated to be about 764 mm of which 35-40% (266 mm) runs off to the oceans (Baumgartner and Reichel, 1975). Assuming no secular trends in the annual mean global soil moisture, this suggests that the annual and global mean evaporation from the land surfaces alone is more than 60% of the annual and global mean precipitation over the land. The percentage is even higher during the local summer for several regions. This suggests that the evaporation from the land surfaces is a very important component of the global water budget and hydrological cycle. However, it does not necessarily follow that the water evaporated from the land is important in determining the rainfall over the land. For example, all the water evporated from the land could be advected away to the oceans before it recondenses and rains. In that case, it will affect the moisture budget and evaporation only over the oceans, which in turn will, of course, affect the moisture supply for rainfall over the land. In order that the evaporation from the land affects the rainfall over the land, it is necessary that the prevailing dynamical circulation be such that the land evaporated moisture recondenses and falls as rain before being advected away. That will depend upon the geographical location of the region under consideration, the prevailing advective velocity, the structure and intensity of the convergence field, and the vertical distribution of moist static energy which determines the nature of moist convection.

The role of soil moisture is twofold. First, it determines the rate of evaporation and, therefore, the moisture supply, and second, it influences the heating of the ground which determines the sensible heat flux and affects the dynamical circulation by generation or dissipation of heat lows. The interaction between the heat lows, generated by solar heating of the ground in the absence of soil moisture, and associated circulation and rainfall is further complicated by the fact that the maintenance and the intensification of the low pressure areas is largely influenced by the latent heat of condensation. For example, if the soil is saturated with water, and the evaporation is equal to the potential evapotranspiration, there will be maximum possible supply of moisture to the atmosphere. Whether increased evaporation will also increase the rainfall will depend upon the structure of dynamical circulation and prevailing flow patterns. If the rate at which the moist static energy ($c_pT + gz + Lq$) is advected away from the region is larger than its accumulation rate, it will not lead to any increase in the rainfall. For the other extreme situation, when the soil is completely dry, and there is no evaporation from the land, there may be a reduction in the rainfall due to reduced evaporation. However, if the heating of the land produces intense low pressure areas which can converge moisture from the surrounding oceans, the rainfall may not necessarily decrease, and if the convergence of moisture is large enough rainfall may even increase. The mechanism will cease to operate, however, once the rain starts falling because the soil will not be dry anymore.

Since the net diabatic heating of a vertical atmospheric column is maximum over the tropical land masses (Figures 7a and 7b), it is quite likely that small fractional changes in these tropical asymmetric heat sources could produce considerable changes in the planetary scale circulations of the tropical as well as the extra-tropical atmosphere. Therefore, in spite of relatively smaller earth surface area being covered by land, soil moisture effects could be as important as SST anomaly effects. It should be noted, however, that the soil moisture effects strongly depend upon the season and latitude because during the winter season in high latitudes, solar radiation reaching the ground is not large enough to be important for surface energy budget. We summarize here the results of two numerical experiments, carried out by Shukla and Mintz (1982) which have demonstrated that evaporation from land can significantly affect the rainfall over land.

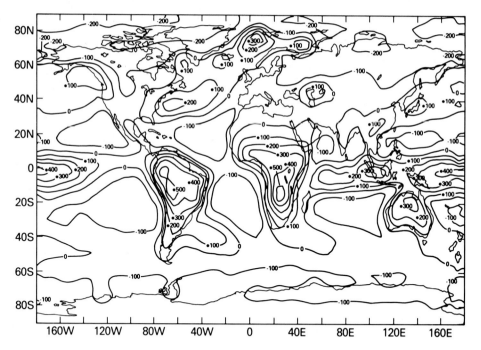

Figure 7a. Vertically integrated net diabatic heating (cal cm^{-2} day^{-1}) of the atmosphere during winter, as simulated by the GLAS climate model

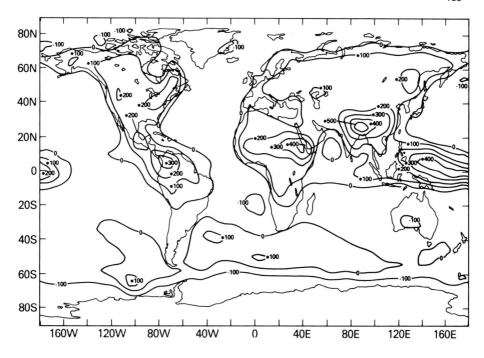

Figure 7b. Vertically integrated net diabatic heating (cal cm^{-2} day^{-1}) of the atmosphere during summer, as simulated by the GLAS climate model

2.2.1. Influence of global dry-soil and wet-soil on atmospheric circulation

We have carried out two 60 day integrations of the GLAS climate model: in one case, there is no evaporation from the land surface ("dry-soil" case) and in the other case the evaporation from the land is equal to the model calculated potential evapotranspiration ("wet-soil" case). These two cases are qualitatively similar to no vegetation and completely vegetated earth surface. For simplicity of interpretation of the results, albedo of the soil was not altered for the two experiments.

Figure 7c shows the global maps of mean July rainfall difference (dry soil-wet soil). Over most of the continental regions, with the exception of the

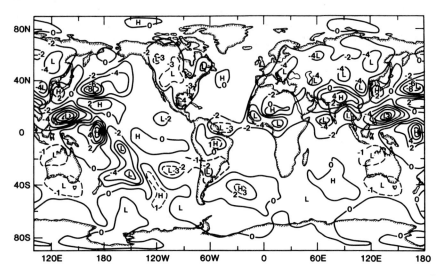

Figure 7c. Rainfall difference (mm/day) between two model simulations (dry soil-wet soil) for the month of July

major monsoon regions, July rainfall for dry-soil case has decreased (by about 40-50%) compared to the wet-soil case. This shows that the evaporation from the land is an important component of the rainfall over the land. The exception over India is even more interesting because the solar heating of land for dry-soil produced such intense low pressure and convergence that the loss of moisture from land evaporation was more than compensated by the increased moisture flux convergence from the neighboring oceanic region. The increased moisture flux convergence leads to increased heating of the vertical air column due to the latent heat of condensation which maintains and intensifies the surface low. This happened prominently for the unique monsoonal circulation over India for which oceanic moisture was brought in from the ocean by the monsoon current. This could also occur in several other regions with monsoonal flow patterns but perhaps the coarse grid resolution of the model could not resolve the local influence over other regions.

Figure 7d shows the ground temperature difference (dry soil-wet soil) for July. Ground temperatures for dry-soil are warmer by more than 20°-30°C. For

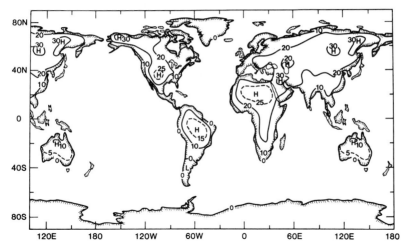

Figure 7d. Ground temperature difference (°C) between two model simulations (dry soil-wet soil) for the month of July

the dry-soil case, most of the radiation energy goes to heat the ground and to increase the sensible heat flux, whereas, for the wet-soil case, most of the radiation energy goes to evaporate the water which can later release the latent heat of condensation and heat the vertical air-column.

Figure 7e shows the surface pressure difference (dry soil-wet soil) for July. There are intense low pressure areas over the continents for dry-soil and the mass removed from over the land is found to produce high pressure areas over the oceans. The location and the intensity of the high pressure areas over the oceans are determined by the nature of dynamical circulations, the most important of which are the mid-latitude stationary wave response to a highly anomalous diabatic forcing, and the modified Hadley and Walker circulations due to changes in the intensity and location of the tropical heat sources.

We have also calculated the natural variability (not shown) of the GLAS model and it should suffice to say that the changes shown here for dry-soil and wet-soil are far too large to be confused with the model variability due to internal dynamics.

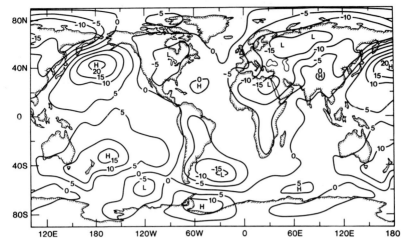

Figure 7e. Surface pressure difference (mb) between two model simulations (dry soil-wet soil) for the month of July

Since the spatially and temporally averaged rainfall over any region depends upon evaporation (which depends upon soil moisture), and vertical distribution of the moisture flux convergence (which depends upon the nature of dynamical circulation, which, in turn is determined by the direct heating of the ground and vertical distribution of latent heat of condensation), it is not possible to establish simple universal relationships between soil moisture, atmospheric circulation and rainfall. Such relationships strongly depend upon the dynamical circulation in the region under consideration. If the moisture over land is rapidly advected away to oceans, it is unlikely that the local evaporation from land will be an important contributor to the rainfall. On the other hand, the intensity and life cycle of a tropical disturbance which moves over land can strongly depend upon the wetness of the ground. It is therefore necessary to utilize realistic physical models of the earth-atmosphere system to determine the sensitivity of climate to fluctuations of soil moisture. The experiments reported in this study are extreme examples to highlight the maximum bounds of impact and to show that the importance of soil moisture and vegetation for rainfall over any particular area should be calculated by including the combined effects of dynamical circulation and orographic and oceanic forcing. It has also been reported that soil moisture boundary forcing can be an important factor for medium range (5-15 day) forecasting of rainfall and circulation. (See Mintz (1982) for a review of several numerical experiments.)

2.3 Surface Albedo.

Charney et al. (1977) had suggested that changes in the surface albedo can produce significant changes in the local rainfall and atmospheric circulation. They pointed out that this effect can be especially important in the desert margin regions of subtropics. Charney et al. used an earlier version of the GLAS model in which the surface albedo was increased by 30% for the Sahel, the Thar Desert (India), and the western Great Plains of the U.S. They found that precipitation over the albedo anomaly regions was reduced by 10-25% within the first 30 days. An increase in the surface albedo reduced the solar radiation reaching the ground, which in turn reduced evaporation from the ground. These reductions in evaporation and cloudiness increased the solar radiation reaching the ground, thus partly compensating for reduction in solar radiation due to increased albedo. However, since reduction in the cloudiness also caused reduction in the long wave radiation emitted back to the surface from the cloud base, there was a net reduction in the total radiative energy coming to the ground. This caused a net reduction in evaporation, cloudiness and precipitation. Since these subtropical regions were not affected by large advective effects, the local changes in radiative and latent heating were accompanied by dynamical circulations which produced descending motion over the albedo anomaly regions. As it was pointed out by Charney et al. (1977), the net effect on the atmospheric circulation due to changes in surface albedo depends upon the relative magnitude of the time scale for advecting the moist static energy away from the region and the time scale for its generation by evaporation and convergence. If the net effect of the change in surface albedo is to reduce the sensible and latent heating of the air, it will either decrease the low level convergence and ascending motion or increase the low level divergence and descending motion. The change in the vertical velocity field in either case will dry the middle troposphere and further reduce precipitation. These effects may not operate in areas of large scale moisture convergence or strong dynamical instabilities; however, these experiments suggested that it is quite important to give a realistic prescription of surface albedo for prediction of monthly and seasonal atmospheric anomalies.

These experiments have been recently repeated by Sud and Fennessy (1982) using the present version of the GLAS climate model which has better parameterizations for evaporation and sensible heat fluxes. Sud and Fennessy have found that the results of experiments with the present model support all the conclusions of the earlier studies by Charney et al. An increase in albedo produced a systematic decrease in rainfall over Sahel and Thar Desert. The only exception occurred for western Great Plains where the change was not large enough to be distinguished from the day-to-day variability.

2.3.1. Effect of change in albedo over northeast Brazil.

In addition to the areas chosen by Charney et al., Sud and Fennessy also examined the effect of change in surface albedo over northeast Brazil. Between 4°S and 24°S, and 32°W and 47.5°W, surface albedo for 10 model grid points was increased from about 9% to 30%. Figure 7f shows 5 day averages of model simulated total rainfall over northeast Brazil for the anomaly (increased albedo) and the control run. It is seen that the rainfall for the first 20 days decreases due to an increase in albedo. For a 5 day period from day 20-25 rainfall in the anomaly case is more than the control. But again, from day 25-45 rainfall for the anomaly case is considerably less than that for the control case. Since rainfall over northeast Brazil is sometimes affected by the penetration of Southern Hemisphere mid-latitude disturbances, the changes during day 20-25 could be attributed to extratropical variability.

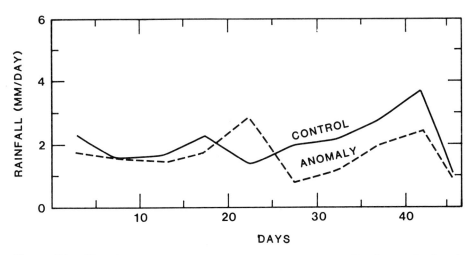

Figure 7f. Five day average rainfall over northeast Brazil, for control and anomaly run

2.4 Snow Cover

About a century ago, Blanford (1884) observed that, "The excessive winter and spring snowfall in the Himalayas is prejudicial to the subsequent monsoon rainfall in India." These observations were later substantiated by Walker (1910). A persistent anomaly of snow cover can affect the meridional temperature gradient and, therefore, vertical shear of the large scale flow. The monsoon circulation is characterized by reversal of the normal temperature gradient between the equator and 30°N, i.e. during the Asiatic summer monsoon season, the equator is colder than northern India. An excessive snowfall during the previous winter and spring season can delay the build-up of the monsoonal temperature gradients because most of the solar energy will be utilized for evaporating the snow or for evaporating the soil moisture due to excessive snow. A weaker meridional temperature gradient between equator and 30°N can give rise to a delayed and weaker monsoon circulation. No systematic numerical experiments have been carried out with global GCMs to determine the physical mechanisms that can affect the atmospheric circulation associated with excessive snowfall. One reason for lack of such numerical experiments is perhaps the requirement of a rather long time integration (from one season to the other) of a global general circulation model. It requires a realistic treatment of albedo as well as surface hydrology because a large and deep snow cover during winter and spring can keep the soil wet for a longer time in the coming summer, and this effect must be treated accurately. It is only recently that general circulation models have shown some success in simulating the seasonal cycle of the atmosphere, and it is hoped that more systematic studies of various snow cover feedback mechanisms will be carried out in the coming years.

Hahn and Shukla (1976) found an apparent relationship between Eurasian snow cover and Indian monsoon rainfall (shown in Figure 8). Large and persistent

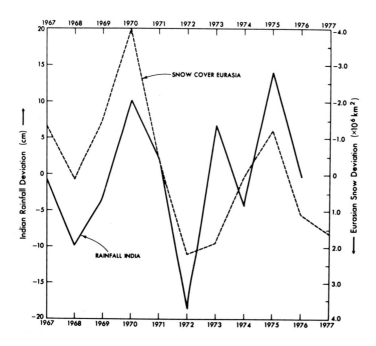

Figure 8.   Area weighted average of percentage departure from normal summer monsoon rainfall over Indian subdivisions (solid line) and the preceding winter snow cover departure over Eurasia south of 52°N

winter snow cover anomalies over Eurasia can produce a colder mid-latitude troposphere in the following spring which can strengthen the upper level anticyclone, slow its northward movement over India, and give rise to delayed and weaker monsoon rainfall. Some recent studies by Yeh et al., (1981) have suggested that Eurasian snow cover anomalies can influence the interannual atmospheric variability of China.

Namias (1962, 1978) has proposed that positive feedback can occur between excessive snow cover on the east coast of North America and quasi-stationary circulations which can be favorable for producing more snowfall. Since the natural variability of the mid-latitude atmosphere is quite large, it is difficult to determine the contribution of snow cover anomalies in producing mid-latitude atmospheric anomalies. The possible physical processes which have the potential to influence the atmospheric circulation due to snow cover anomalies can be briefly mentioned:

(i) An increase in snow cover increases the albedo, and therefore reduces the incoming solar radiation. If there are no other feedbacks, this will produce colder temperatures, and therefore snow cover anomalies will tend to persist for a longer time. This is consistent with the observations of Wiesnet and Matson (1976) who showed that December snow cover for the Northern Hemisphere is a very good predictor of snow cover for the following January through March.

(ii) Excessive snow cover anomalies in the mid-latitudes can produce anomalous diabatic heat sources which in turn can produce anomalous stationary wave patterns which can alter storm tracks and their frequency.

(iii) Persistent snow cover anomalies can change the components of the heat balance of the earth's surface. Even after the snow has melted completely, wet soil can maintain colder surface temperatures for longer periods of time.

(iv) Persistent snow anomalies can produce anomalous meridional temperature gradients and anomalous vertical wind shears. Snow is also a good insulator, and in the presence of deep snow cover, nighttime inversions tend to be much stronger.

2.5 Sea Ice

Although sea ice fluctuations cover a very small fraction of the earth's surface, they can produce important changes in the atmospheric circulation over the polar regions, and for favorable dynamical structure of the large scale flow, these effects could also propagate to middle and subtropical latitudes. The local effect of sea ice anomalies is very large, because it directly changes the heat and moisture supply to the atmosphere. There have been several observational studies to determine possible relationships between sea ice extent and atmospheric circulation anomalies (for a review of these studies, see Walsh (1978), and Walsh and Johnson (1978, 1979)). In a numerical experiment conducted by Herman and Johnson (1979) using the GLAS climate model, it was found that the sea ice anomalies in the Arctic regions not only affected the local circulation but they also produced significant differences in middle and subtropical latitudes. More experimentation with simple models and realistic GCMs are needed to understand the sea ice effects. Since sea ice anomalies, like SST and snow, also change slowly compared to the atmospheric anomalies they can be prescribed from observations.

3. PREDICTABILITY OF THE TROPICAL ATMOSPHERE

The tropical atmosphere is potentially more predictable than the mid-latitudes because its planetary scale circulations are dominated by monsoon circulations which are intrinsically more stable than mid-latitude Rossby regime. Interaction of the large scale overturnings with the tropical disturbances (easterly waves, depressions, cyclones, etc.) is not strong enough to detract from the predictability of the planetary scale circulations. Tropical disturbances are initiated by barotropic-baroclinic instabilities but their main energy source is the latent heat of condensation. Although their growth rate is fast and they are deterministically less predictable, their amplitude equilibration is also quite rapid and with the exception of hurricanes, they attain only moderate intensity. The intensity and geographical locations of the planetary scale circulations are primarily determined by the boundary conditions and not by synoptic scale disturbances. It is reasonable to assume that frequency and tracks of depressions and easterly waves are primarily determined by the location and intensity of the planetary scales and distribution of SST and soil moisture fields. It is highly unlikely that the synoptic scale tropical disturbances, through their interaction with planetary scale disturbances, will drastically alter the character of large scale tropical circulation. This is in marked contrast to the case of mid-latitudes where interaction between synoptic scale instabilities and planetary scale circulations is sufficiently strong so that baroclinically unstable disturbances can make the large scales unpredictable: the mid-latitude circulation consists of a continuum of scales whereas tropical circulation has a clear scale separation in terms of the frequency and the zonal wavenumber.

Charney and Shukla (1981) have suggested that since large scale monsoon circulation is stable with respect to dynamic instabilities, and since boundary conditions exert significant influence on the time averaged monsoon flow, the monsoon circulation is potentially more predictable than the middle latitude circulation. This suggestion was made by examining the variability among the monthly mean (July) circulation of four model runs for which the boundary conditions were kept identical, but the initial conditions were randomly perturbed. It was found that although the observed and the model variabilities were comparible for middle and high latitudes, the variability among the four model runs for the monsoon region was far less than the observed interannual variability of the atmosphere as a whole. This led to the suggestion that part of the remaining variability could be due to the boundary forcings.

We have extended the work of Charney and Shukla (1981) and compared the model variability for climatological and observed SST anomalies. We have carried out 45 day integrations of the GLAS climate model for seven different sets of initial conditions and boundary conditions. In four of these integrations, the climatological global SST was used. In the remaining three, the observed SST for 1973, 1974 and 1975 was used between 0-30°N. Figure 9 shows the plots

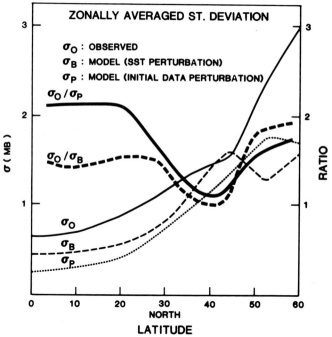

Figure 9. Zonally averaged standard deviation among monthly mean (July) sea level pressure (mb) for 10 years of observations (σ_O, thin solid line); four model runs with variable boundary conditions (σ_B, thin dashed line); and four model runs with identical boundary conditions (σ_P, thin dotted line). Thick solid line and thick dashed line show the ratio σ_O/σ_P and σ_O/σ_B respectively

of zonally averaged values of standard deviations σ_p, σ_B and σ_0, and ratios σ_0/σ_p and σ_0/σ_B as a function of latitude. σ_p is the model standard deviation among predictability integrations (climatological SST and random perturbation in initial conditions), σ_B is the model standard deviation for SST anomalies, and σ_0 is the standard deviation for 10 years of observations. It is seen that, in agreement with the results of Charney and Shukla, the ratio σ_0/σ_p is more than two in the tropics and close to one in the middle latitudes. The new result of this study is that σ_B (variability due to changes in SST boundary conditions) lies nearly halfway between σ_0 and σ_p. This suggests that nearly half of the potentially predictable variability is accounted for by changes in SST between 0-30°N.

These conclusions are further supported by a more comprehensive study by Manabe and Hahn (1981), who integrated the GFDL spectral climate model for 15 years with prescribed but seasonally varying boundary conditions of SST. Figure 10a (reproduced from Manabe and Hahn (1981)) shows the values of zonally averaged standard deviation of 1000 mb geopotential height for a 15 year model run (σ_m) and observations (σ_0). We have calculated the ratio (σ_0/σ_m) from the two curves of Manabe and Hahn and the ratio is also shown in Figure 10a. It is again seen that the ratio (σ_0/σ_m) is about two in the near equatorial regions and reduces to about one in the middle and high latitudes. Figure 10b (which is reproduced from Figure 5.10 of Manabe and Hahn) shows the latitude and height cross sections of interannual variability of geopotential height. It is seen that the ratio between the observed variability and the model variability is more than three in the tropical upper troposphere. It should be pointed out, however, that although SST was not varying from one year to the other for the 15 year model run, the soil moisture and snow cover were still variable during different years, and therefore, it is likely that part of the simulated model variability in the tropics, and perhaps even in the middle latitudes, could be due to the interannual variability of soil moisture and snow cover.

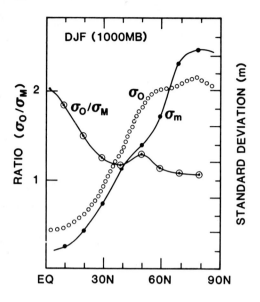

Figure 10a. Zonal means of standard deviation of monthly mean 1000 mb geopotential height (m) for the Dec.-Jan.-Feb. season. Observed distributions are from Oort and Jenne. The ratio of observed and model standard deviation is on the left hand side. (from Manabe and Hahn, 1981)

It is reasonable to conclude that although for short and medium range the tropical atmosphere is less predictable, the time averages (monthly and seasonal means) for the tropics have more potential predictability. Since there is sufficient evidence that tropical heat sources can also influence the middle latitude circulation, it is likely that, even for mid-latitudes, the monthly means could be potentially predictable due to their interaction with low latitudes.

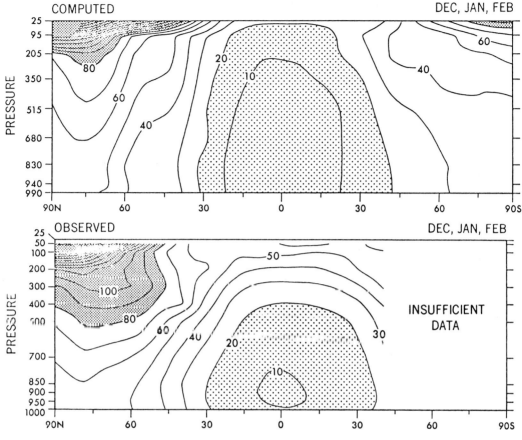

Figure 10b. Zonal mean of the standard deviation of monthly mean geopotential height (m) for the Dec.-Jan.-Feb. season. Top: simulated; Bottom: observed (from Manabe and Hahn, 1981)

4. POTENTIAL PREDICTABILITY OF MONTHLY MEANS AS DEDUCED FROM OBSERVATIONS

As pointed out in the Introduction, the interannual variability of monthly means is determined by complex interactions between the internal dynamics and surface boundary forcings. It is therefore difficult to determine their separate contributions by analysis of observed data. One possible approach is to carry out controlled numerical experiments described earlier, provided that all the boundary forcings can be reasonably prescribed.

Madden (1976) has examined the predictability of monthly mean sea level pressure over the Northern Hemisphere by comparing the variances of the observed monthly means with the natural variability of the monthly means. His study concluded that there are several areas over the Northern Hemisphere for which the ratio of the above two variances is more than one and therefore there is potential predictability. However, in Madden's study, the predictable signal was assumed to be only that part of the variance which is above white noise for more than 96 day time scales. He further assumed that the boundary forcings do not contribute to variability for time scales shorter than 96 days. Since it is well known that changes in boundary forcings can influence atmospheric fluctuations within 96 days, Madden's results should be considered only as the lower bound of the estimate of potential predictability of the monthly means of the atmospheric circulations (Shukla, 1983). Moreover, Madden's study does not address the question of dynamical prediction from an initial state.

Shukla and Gutzler (1983) have carried out an analysis of variance to compare the interannual variability of 500 mb geopotential height among 15 (1963-77) January months, and variability within each January. It can be hypothesized that if variability among the Januaries of different years is significantly larger than that due to the day to day variability within the individual Januaries, the excess variability could be due to boundary forcings. Since daily values within the individual Januaries are not independent of each other, we first calculated the time interval between independent samples which were used to calculate the effective sample size and effective degrees of freedom, and then calculated the natural variability of January means.

Figure 11a shows the ratio of the observed variability among the Januaries of different years and the natural variability. It is seen that the ratio between the two variances is larger than two (in some places, even larger than three) over a substantial part of the Northern Hemisphere. This suggests that

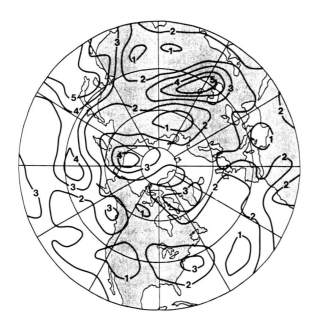

Figure 11a. Ratio of the variances of observed January mean and the natural variability of January mean geopotential height at 500 mb

the boundary forcings play an important role in determining the interannual variability of monthly means. As mentioned earlier, these observational studies also suggest that for time averages (monthly and seasonal means) the tropical atmosphere is potentially more predictable than the mid-latitude atmosphere. This suggestion is further supported by a comparison of autocorrelation functions for the tropics and mid-latitudes. Figure 11b shows the zonally averaged values of autocorrelation calculated from 53 years (1925-77) of monthly mean sea level pressure data over the Northern Hemisphere analyzed by Trenberth and Paolino (1981). It is seen that the autocorrelation drops off much more sharply for mid-latitudes than in the tropics. The tropical spectra are redder than the mid-latitude spectra and therefore potentially more predictable.

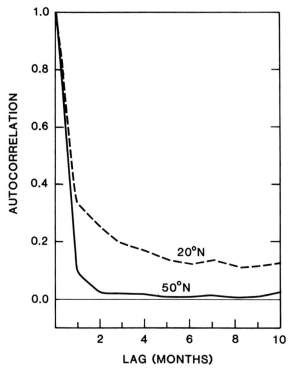

Figure 11b. Zonally averaged autocorrelation at 20°N and 50°N for monthly mean sea level pressure for the period 1925-77

5. CONCLUSIONS

We have briefly reviewed the physical mechanisms through which boundary forcings can influence the variability and predictability of monthly means. We have used a realistic global general circulation model to determine the sensitivity of the model circulation to changes in the boundary conditions of SST, soil moisture and surface albedo. From these studies it can be concluded that the slowly varying boundary forcings have the potential to increase the predictability of time averages. It is further suggested that correct specification of global boundary conditions of SST, soil moisture, surface albedo, sea ice, and snow is necessary for successful dynamical prediction of monthly means.

Anomalies of slowly varying boundary forcings produce anomalous sources of heat and moisture which in turn produce significant anomalies of atmospheric circulation. These effects can be either local in the vicinity of the boundary anomaly or away from the source if the intervening environment is favorable for the propagation of local influences. For example, tropical heat sources can produce significant changes in the extra-tropical circulation.

It should be pointed out that the anomalies of SST or soil moisture are neither necessary nor sufficient to produce changes in the mid-latitude circulation. For example, a warm SST anomaly in tropics can produce a local heat source, but if it occurs within the band of easterlies, its influence cannot propagate beyond the zero wind line. Conversely, even in the absence of SST and soil moisture anomalies, internal dynamics can produce persistent anomalies of precipitation and diabatic heating in the tropics which, under favorable zonal flow, can affect the mid-latitudes.

The main purpose of the two papers (Shukla (1981) and this paper) was to examine and establish a physical basis for dynamical prediction of monthly means. The essential reqiurements for establishing such a basis are to show that: a) fluctuations of monthly means are larger than can be expected due to sampling of day-to-day weather changes; b) there are low frequency planetary scale components of the circulation which remain predictable beyond the limits of synoptic scale predictability; and, c) influences of the slowly varying boundary conditions of SST, soil moisture, snow, sea ice, etc., are large enough to produce significant and detectable changes in the monthly mean circulations. In these two papers we have presented observational and numerical evidence which support the above requirements.

It is hoped that an accurate description of global boundary forcings can be obtained on a near-real time basis so that feasibility of experimental prediction of monthly means can be investigated using dynamical models.

ACKNOWLEDGEMENTS

We are grateful to Yale Mintz and Mike Wallace for reading the manuscript and making many valuable suggestions. We thank Miss Debbie Boyer for typing the manuscript and Miss Laura Rumburg for drafting the figures.

REFERENCES

Baumgartner, H., and E. Reichel, 1975: The world water balance: Mean annual global continental and maritime precipitation, evaporation and runoff. (Elsevier, Amsterdam/Oxford/New York, 179 pp and plates.)

Blanford, H. F., 1884: On the connection of the Himalaya snowfall with dry winds and seasons of droughts in India. Proc. Roy. Soc., London, 37, p. 3.

Charney, J. G., W. J. Quirk, S. Chow, and J. Kornfield, 1977: A comparative study of the effects of albedo change on drought in semi-arid regions. J. Atmos. Sci., 34, 1366.

Charney, J. G., and J. Shukla, 1981: Predictability of monsoons. Monsoon Dynamics, Cambridge University Press, Editors: Sir James Lighthill and R. P. Pearce.

Chen, T. C., and J. Shukla, 1983: Diagnostic analysis and spectral energetics of a blocking event in the GLAS climate model simulation. Mon. Wea. Rev., 111, 3-22.

Hahn, D., and J. Shukla, 1976: An apparent relationship between Eurasian snow cover and Indian monsoon rainfall. J. Atmos. Sci., 33, 2461-2463.

Horel, J. D., and J. M. Wallace, 1981: Planetary-scale atmospheric phenomena associated with the Southern Oscillation. Mon. Wea. Rev., 109, 813-829.

Hoskins, B. J., and D. J. Karoly, 1981: The steady linear response of a spherical atmosphere to thermal and orographic forcing. J. Atmos. Sci., 38, 1179-1196.

Lambert, S. J., and P. E. Merilees, 1978: A study of planetary wave errors in a spectral numerical weather prediction model. Atmos. Ocean, 16, 197-211.

Madden, R. A., 1976: Estimates of the natural variability of time-averaged sea-level pressure. Mon. Wea. Rev., 104, 942-952.

Manabe, S., and D. G. Hahn, 1981: Simulation of atmospheric variability. To be published in Mon. Wea. Rev.

Mintz, Y., 1982: The sensitivity of numerically simulated climates to land-surface boundary conditions. NASA Tech. Memo. 83985, 81 pp.

Moura, A. D., and J. Shukla, 1981: On the dynamics of droughts in northeast Brazil: Observations, theory and numerical experiments with a General Circulation Model. J. Atmos. Sci., 38, 2653-2675.

Namias, J., 1962: Influence of abnormal surface heat sources and sinks on atmospheric behavior, Proceedings of the International Symposium on Numerical Weather Prediction, pp. 615-629, Meteorol. Soc. of Japan, Tokyo.

Namias, J., 1978: Multiple causes of the North American abnormal winter 1967-77. Mon. Wea. Rev., 106, 279-295.

Rasmusson, E., and T. Carpenter, 1982: Variations in tropical sea surface temperature and surface wind fields associated with the Southern Oscillation/El Nino. Mon. Wea. Rev., 110, 354-384.

Shukla, J., 1975: Effect of Arabian sea-surface temperature anomaly on Indian summer monsoon: A numerical experiment with the GFDL model. J. Atmos. Sci., 32, 503-511.

Shukla, J., and B. Bangaru, 1979: Effect of a Pacific sea surface temperature anomaly on the circulation over North America. GARP Publication Series No. 22, 501-518.

Shukla, J., 1981: Dynamical predictability of monthly means. J. Atmos. Sci., 38, 2547-2572.

Shukla, J., 1983: Comments on natural variability and predictability. Mon. Wea. Rev., 111.

Shukla, J., and D. S. Gutzler, 1983: Interannual variability and predictability of 500 mb geopotential heights over the Northern Hemisphere. Mon. Wea. Rev., 111.

Shukla, J., and R. Lindzen, 1981: Stationary waves and deterministic predictability. Presented at the Third Conference on Atmospheric and Oceanic Waves and Stability, January 19-23, 1981, San Diego, California.

Shukla, J., and Y. Mintz, 1982: The influence of land-surface evapotranspiration on the earth's climate. Science, 214, 1498-1501.

Shukla, J., and J. M. Wallace, 1983: Numerical simulation of the atmospheric response to equatorial Pacific sea surface temperature anomalies. J. Atmos. Sci., 40.

Sud, Y., and M. Fennessy, 1982: A numerical simulation study of the influence of surface-albedo on July circulation in semi-arid regions using the GLAS GCM. J. of Climatology, 2, 105-125.

Trenberth K. E., and D. A. Paolino, 1981: Characteristic patterns of variability of sea-level pressure in the Northern Hemisphere. Mon. Wea. Rev., 109, 1169-1189.

Walker, G. R., 1910: Correlations in seasonal variations of weather II. Mem. Indian Meteor. Dept., 21, 22-45

Walsh, J. E., 1978: Temporal and spatial scales of the Arctic circulation. Mon. Wea. Rev., 106, 1532-1544.

Walsh, J. E., and C. M. Johnson, 1979: An analysis of Arctic Sea ice fluctuations, 1953-77. J. Phys. Oceanogr., 9, 580-591.

Walsh, J. E., and C. M. Johnson, 1979: Interannual atmospheric variability and associated fluctuations in Arctic Sea ice extent. J. Geophys. Res., 84, 6915-6928.

Washington, W. M., R. M. Chervin, and G. V. Rao, 1977: Effects of a variety of Indian Ocean surface temperature anomaly patterns on the summer monsoon circulation: Experiments with the NCAR General Circulation Mode. Pageoph, 115, 1335-1356.

Webster, P. J., 1981: Mechanisms determining the atmospheric response to sea surface temperature anomalies. J. Atmos. Sci., 38, 554-571.

Wiesnet, D. R., and M. Matson, 1976: A possible forecasting technique for winter snow cover in the Northern Hemisphere and Eurasia. Mon. Wea. Rev., 104, 828-835.

Yeh, T.-C., X.-S. Chen, and C.-B. Fu, 1981: The time-lag feedback processes of large-scale precipitation on the atmospheric circulation and climate--the air-land interaction. (pre-published manuscript)

Statistical Methods for the Verification of Long and Short Range Forecasts

C. E. LEITH[*]

1. INTRODUCTION

Weather forecasts are often accurate but never exact. Forecasting must be treated in part therefore as a random process with errors distributed according to some probability distribution. It is important for us to quantify these error distributions not only as a guide to users of the forecasts but also as a measure of progress in the development of improved forecasting methods. A general theory of observation and prediction error was first formulated by Gauss (1809) for orbit calculations in celestial mechanics and has since been applied in many fields of science and technology.

In these lecture notes I shall show how these general ideas have been applied to an analysis of forecasting error. I have two general goals. The first is to develop a calculus of error variance from which a simple forecast error budget may be constructed. This is done in Sections 3 and 4. The second is to provide the basis for judging the significance of any change in error statistics in tests of proposed improvements in an observing and forecasting system. This is done in Section 5. In Section 2 I summarize the general definitions of random variables and vectors which I apply to random errors in Section 3.

2. Random variables and vectors

2.1 Random scalars

In common usage the adjective random means uncertain, unknown, or unpredictable, and a random variable would be considered in some way ill-defined. In probability theory a <u>random variable</u> is defined as a variable which takes on different values with a specified proba-

---

[*] On leave from the National Center for Atmospheric Research which is sponsored by the National Science Foundation.

bility. Thus associated with a real random variable x is a probability function P(x) such that the probability that $x \leq x_0$ is $P(x_0)$. If P(x) is differentiable then there is an associated <u>probability density distribution</u> p(x) such that $dP(x) = p(x)dx$. Without much loss in generality I shall consider that this is the case in these notes.

A random variable may be characterized by a hypothetical <u>ensemble</u> of an infinite number of possible and equally likely members each labeled by a value of x. The probability density p(x) is then a measure of the number density of members at x when they are sorted and distributed along an x-axis.

A probability density p(x) must clearly be normalized,

$$\int_{-\infty}^{\infty} p(x)\,dx = P(\infty) = 1 \qquad (2.1)$$

and everywhere non-negative,

$$p(x) \geq 0. \qquad (2.2)$$

but is otherwise unconstrained. For an arbitrary function f(x) of a random variable, which attaches to a member of the ensemble labeled by x a functional value f, we denote the <u>average</u> of f over the ensemble as

$$<f> = \int_{-\infty}^{\infty} f(x)\,p(x)\,dx \qquad (2.3)$$

Only if the integral converges does the average of f exist as a finite number.

<u>Moments</u> are defined as averages of powers, thus the nth moment is given by

$$<x^n> = \int_{-\infty}^{\infty} x^n p(x)\,dx. \qquad (2.4)$$

The zeroth moment is identically equal to unity by the normalization condition (2.1). We shall be primarily interested in the first and second moments. The first moment $<x>$ is called the <u>mean</u> of the distribution p(x). Averages such as the mean are not random variables but <u>sharp variables</u> or constants independent of x. As for the zeroth moment, the average of a sharp variable is equal to itself. The averaging operation is clearly linear, so that if we define a new random variable as

$$y = x - <x> \qquad (2.5)$$

its mean will vanish since

$$\langle y \rangle = \langle x \rangle - \langle \langle x \rangle \rangle = \langle x \rangle - \langle x \rangle = 0. \tag{2.6}$$

Higher order central moments or moments about the mean are defined in terms of this shifted variable. The <u>variance</u>, X, is the second central moment,

$$X = \langle (x-\langle x \rangle)^2 \rangle = \langle x^2 \rangle - \langle x \rangle^2 \tag{2.7}$$

I shall generally denote the variance as here by capitalization.

In his analysis Gauss (1820) introduced for simplicity the <u>normal</u> (or Gaussian) <u>distribution</u>

$$p(x) = (2\pi)^{-1/2} X^{-1/2} \exp\left[-(\tfrac{1}{2}) X^{-1} (x-\langle x \rangle)^2\right] \tag{2.8}$$

as a plausible distribution for random errors. It depends on only two parameters $\langle x \rangle$ and X which are, as indicated, its mean and variance. We shall follow the lead of Gauss and describe a theory of errors in terms of first and second moments only. We shall also assume that we are dealing with normal probability distributions.

2.2 Random vectors

A <u>random vector</u> is characterized by an ensemble of vectors $\underline{x}$ in an n-dimensional vector space with an associated probability density distribution which is nonnegative

$$p(\underline{x}) \geq 0. \tag{2.9}$$

and normalized

$$\int p(\underline{x}) d\underline{x} = 1 \tag{2.10}$$

A random vector is generally more than a collection of random components. If the components are <u>independent</u> random variables with distributions $p_1(x_1), p_2(x_2), \ldots p_n(x_n)$ then

$$p(\underline{x}) = p_1(x_1) p_2(x_2) \cdots p_n(x_n), \tag{2.11}$$

but in general $p(\underline{x})$ can not be so factored and induces probabilistic relations between the components.

Averages and moments are defined as generalizations of their definition for random variables. In particular, the mean is a vector

$$\langle \underline{x} \rangle = \int \underline{x}\, p(\underline{x})\, d\underline{x} \tag{2.12}$$

For notational convenience a random vector $\underline{x}$ will be treated as a column vector in component form with $\underline{x}^*$ indicating the transposed row vector. Then $\underline{x}\underline{x}^*$, in usual matrix multiplication notation, is an n×n square matrix. The <u>covariance matrix</u> is defined as the central second moment

$$\underline{X} = \langle (\underline{x}-\langle \underline{x}\rangle)(\underline{x}-\langle \underline{x}\rangle)^* \rangle = \langle \underline{x}\underline{x}^* \rangle - \langle \underline{x}\rangle\langle \underline{x}^*\rangle \tag{2.13}$$

Its diagonal elements are variances of the associated components; its nondiagonal elements are covariances of the associated pairs of components. An important property of a covariance matrix is that it is non-negative definite, which means that for any (sharp) vector $\underline{a}$ the scalar, $\underline{a}^*\underline{X}\underline{a}$ is nonnegative. To see this let $\underline{y} = \underline{x} - \langle \underline{x}\rangle$ so that, since $\underline{a}^*\underline{y} = \underline{y}^*\underline{a}$ is a random scalar, we have

$$\underline{a}^*\underline{X}\underline{a} = \underline{a}^*\langle \underline{y}\underline{y}^*\rangle \underline{a} = \langle (\underline{a}^*\underline{y})(\underline{y}^*\underline{a})\rangle = \langle (\underline{a}^*\underline{y})^2\rangle \geq 0. \tag{2.14}$$

By a suitable linear transformation of coordinates the covariance matrix $\underline{X}$ may be diagonalized, thus displaying its eigenvalues as variances on the diagonal of a matrix whose non-diagonal covariance elements vanish. These eigenvalues are nonnegative, but some might vanish. The associated eigenvector would in that case be nonrandom or sharp. We shall assume that such a degeneracy does not occur in our applications and thus that $\underline{X}$ is positive definite, i.e. that $\underline{a}^*\underline{X}\underline{a}$ is strictly positive. We shall then feel free to invert $\underline{X}$ at will.

By letting the dimensionality of the random vectors become infinite it is possible to generalize the definitions to apply to random functions or field. We shall not do this since the meterological fields of interest for weather forecasting must be represented for computing purposes in terms of a large but finite member of degrees of freedom. But this serves as a reminder that we are interested in random vectors whose components may consist of all the spectral components in a prediction model and may have a dimension of order 10^5.

3. Probabilistic measures of error

3.1 Scalar error

Let us suppose that we measure using an imperfect technique a physical quantity, whose true value, is x_t. Owing to imperfections the measured value x_m will differ from x_t by a measurement error $e = x_m - x_t$. We consider the measurement to select one member out of an ensemble of equally likely possible errors, and thus the error e is a random variable with some probability density p(e). We shall assume that the first two moments of p(e) are known from experience such as provided by repeated independent applications of the measurement technique to a known standard. The first moment $<e>$ is the <u>mean error</u> or <u>bias</u> which by recalibration we may assume to vanish. The second central moment, $E = <e^2>$ is the <u>mean square error</u> or <u>error variance</u> which, following Gauss, we take to be the primary probabilistic measure of error. Clearly the smaller is E the more accurate is the measurement technique. We shall define the accuracy $A = E^{-1}$ as the inverse of E.

The random error e induces randomness in the measured value

$$x_m = x_t + e \tag{3.1}$$

even though x_t is sharp. Over an infinite ensemble of measurements the mean of x_m is

$$<x_m> = x_t + <e> = x_t \tag{3.2}$$

and its variance is

$$E + <e>^2 = E. \tag{3.3}$$

Suppose in addition that x_t is a random variable with mean 0 and variance X_t and that the error e is independent of x_t. Then the variance X_m of measured values is given by

$$X_m = <x_m^2> = <(x_t+e)^2> = X_t + E \geq X_t \tag{3.4}$$

since, in this case, $<x_t e> = 0$. Note that this result is unchanged by a change in sign of e. Similarly if there are two independent errors e_1 and e_2 in a measurement then

$$x_m = x_t + e_1 + e_2 \tag{3.5}$$

and

$$X_m = X_t + E_1 + E_2 \tag{3.6}$$

It is this additive property of independent sources of variance that makes variance such a useful simple measure of error.

Another common error measure is the <u>standard error</u> or <u>root mean square</u> (<u>rms</u>) error defined in general as $<e^2>^{1/2}$. This is a measure of the width of the error probability distribution in the same physical units as x_m. The standard error does not, however, have the additive property of error variance.

Accuracy can be additive also. Consider two measurements of x_t made by separate, independent, and perhaps differing measurement techniques giving the pair of results

$$x_{m1} = x_t + e_1$$
$$x_{m2} = x_t + e_2 \tag{3.7}$$

We assume that each measurement is unbiased, but that they may have different error variances E_1 and E_2. One expects that some weighted combination of the two measured values

$$x_m = q x_{m1} + (1-q) x_{m2} \tag{3.8}$$

should be more accurate than either by itself. This is, in fact, the case in a probabilistic sense. Clearly the combined error

$$e = q e_1 + (1-q) e_2 \tag{3.9}$$

has a variance

$$E = \left< \left[q e_1 + (1-q) e_2 \right]^2 \right> = q^2 E_1 + (1-q)^2 E_2 \tag{3.10}$$

which may be minimized by finding a value of q such that

$$dE/dq = 2q E_1 - 2(1-q) E_2 = 0 \tag{3.11}$$

Eq. (3.11) is satisfied by

$$q = \frac{E_2}{E_1 + E_2} = \frac{A_1}{A_1 + A_2}, \quad 1 - q = \frac{A_2}{A_1 + A_2} \tag{3.12}$$

where $A_1 = E_1^{-1}$, $A_2 = E_2^{-1}$ are the respective accuracies. Note that the weights are proportional to the corresponding accuracies and that this is indeed a minimum in E since

$$d^2E/dq^2 = 2(E_1+E_2) > 0. \tag{3.13}$$

The value of E at the minimum is given by

$$E = (A_1+A_2)^{-2} \left[A_1^2 E_1 + A_2^2 E_2 \right]$$
$$= (A_1+A_2)^{-1} \tag{3.14}$$

The final accuracy

$$A = E^{-1} = A_1 + A_2 \tag{3.15}$$

is the sum of the two contributing accuracies and therefore greater than either.

A special case of some interest is for an independent repetition of the same measurement technique whence $E_2 = E_1$, $A_2 = A_1$, and the two weights are equally $1/2$. Then $A = 2A_1$ and $E = \frac{1}{2}E_1$. This can be extended to many repetitions, thus for n independent measurements the best estimate of x_t is an equally weighted average of the x_{mi}'s and its accuracy is n times greater than that of a single measurement.

In summary, independent sources of error lead to additive error variances, but independent sources of information lead to additive accuracies.

3.2 Error vectors

For an imperfect measurement of a vector $\underline{x}_t$ we have

$$\underline{x}_m = \underline{x}_t + \underline{e} \tag{3.16}$$

where $\underline{e}$ is a random error vector. The extension of error measures to random vectors is fairly straightforward. We may again without much loss of generality set the mean error, now a vector, to zero, thus $<\underline{e}> = \underline{0}$, where $\underline{0}$ is a vector all of whose components are zero. The error is then characterized by a positive definite <u>error covariance matrix</u>

$$\underline{E} = <\underline{e}\,\underline{e}^*> \tag{3.17}$$

whose matrix inverse

$$\underline{A} = \underline{E}^{-1} \tag{3.18}$$

serves as a positive definite <u>accuracy matrix</u>.

In the case that all error covariance and accuracy matrices under discussion commute they all have a common set of eigenvectors and can all be diagonalized by the same transformation. Each error eigenmode then is independent of all others, and the scalar algebraic manipulations of adding error variance or accuracies carry through separately.

In particular, two independent sources of error with covariance matrices $\underline{E}_1$ and $\underline{E}_2$ contribute a total error with covariance matrix $\underline{E} = \underline{E}_1 + \underline{E}_2$. Two independent sources of information with accuracy matrices $\underline{A}_1$ and $\underline{A}_2$ provide a total accuracy $\underline{A} = \underline{A}_1 + \underline{A}_2$ when combined with matrix weights

$$\underline{Q}_1 = \underline{A}^{-1} \underline{A}_1, \quad \underline{Q}_2 = \underline{A}^{-1} \underline{A}_2 \tag{3.19}$$

so that

$$\underline{x} = \underline{A}^{-1} \underline{A}_1 \underline{x}_1 + \underline{A}^{-1} \underline{A}_2 \underline{x}_2 \tag{3.20}$$

In case the matrices $\underline{E}_1$ and $\underline{E}_2$ do not commute the summing of error covariances remains valid, but the solution to the problem of combining information is not so clear. The use of Eqns (3.19) and (3.20) appears however, to remain plausible (Leith 1975).

In the case that $\underline{x}_t$ itself is random, with say $<\underline{x}_t> = 0$ and covariance matrix $\underline{X}_t = <\underline{x}_t \underline{x}_t^*>$, then this knowledge can serve as an independent source of information to improve the estimate $\underline{x}_m$. We seek, in this case, a regression matrix $\underline{R}$ such that

$$\underline{x}_t = \underline{R} \underline{x}_m + \underline{f} \tag{3.21}$$

This is a sort of inversion of Eqn (3.16). Here

$$\hat{\underline{x}}_t = \underline{R} \underline{x}_m \tag{3.22}$$

is a best estimate of $\underline{x}_t$ with error covariance

$$\underline{F} = <\underline{f}\ \underline{f}^*> \tag{3.23}$$

if $\underline{R}$ is chosen such that

$$<\underline{f}\ \underline{x}_m^*> = \underline{0} \tag{3.24}$$

for then $\underline{x}_m$ provides no further information about $\underline{f}$. We have from

Eqn (3.21)

$$<\underline{x}_t \underline{x}_m^*> = \underline{R}<\underline{x}_m \underline{x}_m^*> = \underline{R}\ \underline{X}_m \qquad (3.25)$$

and from Eqn (3.16)

$$<\underline{x}_t \underline{x}_m^*> = <\underline{x}_t \underline{x}_t^*> = \underline{X}_t \qquad (3.26)$$

so that with the use of Eqn (3.4) we find

$$\underline{R} = \underline{X}_t\ \underline{X}_m^{-1} = \underline{X}_t(\underline{X}_t+\underline{E})^{-1}$$

$$= [(\underline{X}_t+\underline{E})\underline{X}_t^{-1}]^{-1} = [\underline{I}+\underline{E}\ \underline{X}_t^{-1}]^{-1}$$

$$= [\underline{E}(\underline{E}^{-1}+\underline{X}_t^{-1})]^{-1} = [\underline{E}^{-1}+\underline{X}_t^{-1}]^{-1}\underline{E}^{-1}$$

$$= (\underline{A}+\underline{A}_t)^{-1}\underline{A} \qquad (3.27)$$

where $\underline{A} = \underline{E}^{-1}$ and $\underline{A}_t = \underline{X}_t^{-1}$.
The final error covariance is

$$\underline{F} = <\underline{f}(\underline{x}_t-\underline{R}\ \underline{x}_m)^*> = <\underline{f}\ \underline{x}_t^*>$$

$$= <(\underline{x}_t-\underline{R}\ \underline{x}_m)\underline{x}_t^*> = \underline{X}_t - \underline{R}<\underline{x}_m\ \underline{x}_t^*>$$

$$= \underline{X}_t - \underline{R}\ \underline{X}_t = [\underline{I} - (\underline{A}+\underline{A}_t)^{-1}\underline{A}]\underline{X}_t$$

$$= (\underline{A}+\underline{A}_t)^{-1}\underline{A}_t\ \underline{X}_t = (\underline{A}+\underline{A}_t)^{-1} \qquad (3.28)$$

Standard regression analysis leads thus to the expected result. $\underline{R}$ is the normalized weight for $\underline{x}_m$ in the combination, $\underline{I}-\underline{R}$ is the normalized weight for $<\underline{x}_t> = \underline{0}$, and the final accuracy is $\underline{A} + \underline{A}_t$. No assumption has been made that $\underline{E}$ and $\underline{X}_t$ or $\underline{A}$ and $\underline{A}_t$ commute.

3. Forecasting errors

I have described the error analysis in terms of measurement error, but it is equally valid for forecasting error. In that case $\underline{x}_m = \underline{x}_f$ represents the forecast atmosphere state vector in a model and $\underline{x}_t$ represents the state vector in the model that most closely describes the true state of the atmosphere at that time. The error ensemble is based hypothetically on an infinite number of forecasts for an infinite ensemble of atmospheric states. This ensemble of states $\underline{x}_t$ can be considered as the climate ensemble and it is convenient to set $<\underline{x}_t> = 0$ by defining the state vector components as anomalies from the climate mean. The regression analysis that utilizes information about the climate covariance matrix $\underline{X}_t$ is an example of "statistically optimal analysis" when applied to observations or of "optimal filtering" when applied to forecasts.

One practical problem in utilizing random error vector analysis becomes clear immediately. The number of different components in an error covariance matrix $\underline{E}$ is $N(N+1)/2$ where N is the dimension of the model state vector. For N of order 10^5 this is a tremendous amount of information. In practice therefore one usually reduces the measure of error back to a scalar variance quantity by some combination of selection and averaging over diagonal elements. Such a procedure can be strictly justified only if all covariance matrices involved are multiples of a common matrix. In the next section I shall develop simple error budget equations based on this kind of a reduction.

4. Simple prediction error budget equations

4.1 Introduction

It is known from many theoretical predictability studies that even for a perfect prediction model any initial error, no matter how small in scale and amplitude, will progressively contaminate increasingly larger scales and grow with an rms error doubling time of about 2 1/2 days. Prediction models, however, are not perfect and introduce an additional source of error growth. A simple error-growth equation (Leith, 1978) has been used to describe the initial growth of small errors and to distinguish model-induced error sources from the inherent growth of initial analysis errors. Although the equation is based on rather crude assumptions, it seems to provide a consistent fit to observed error growths during the first day or so of prediction. It has recently been applied with some success to the operational ECMWF prediction model by Bengtsson (1981).

In this section the equation will be summarized and extended to the calculation of the error budget for a standard data assimilation procedure. This extension provides an estimate of the relative impact of model errors on the equilibrium error level of the final analysis with assimilation. The saturation effects of climate variance on error growth at late times will also be discussed.

4.2 Error variance

In dealing with error budgets, it is far more natural to use mean square error or error variance E rather than the commonly used root mean square (rms) error. If, for example, a particular determination is afflicted by two independent errors with variances E_1 and E_2, then the resulting error variance is the simple sum $E = E_1 + E_2$. If, on the contrary, two independent determinations with error variances E_1 and E_2 are combined into a better final determination, then the inverses, which measure accuracy or information content, are summed; $E^{-1} = E_1^{-1} + E_2^{-1}$. We shall use both of these general statistical principles which were developed in Section 3.

The use of a single number E to describe the error variance of an atmospheric state ignores details of error distributions over space or over spatial scales. This is less serious for compositing error than for compositing information, but in either case an implicit assumption is made that all distributions are similar.

4.3 Error growth equation

The simple error growth equation (Leith, 1978) is

$$\dot{E} = \alpha E + S \qquad (4.1)$$

The term αE describes the inherent tendency for error to grow owing to the unstable nature of atmospheric dynamics. The rms error doubling time of 2 1/2 days given by predictability theory translates into an error variance doubling time of 1.25 days and a value of $\alpha = 0.5545$ day^{-1}. The term S describes the model error source rate, which is model dependent and can be empirically determined by fitting observed error growth values.

Analysis error variance, which includes observation errors, will be denoted by E_0 and provides an initial value for the integration of Eq. (4.1) with the result

$$E(t) = E_0 + (E_0 + S/\alpha)[\exp(\alpha t) - 1] \qquad (4.2)$$

It is convenient to replace the time variable t with the pseudo-time variable

$$\tau = \frac{1}{\alpha}[\exp(\alpha t) - 1] \qquad (4.3)$$

in terms of which the error growth is linear. The perceived error variance involves a verification against a later analysis, and this contributes an additional term E_0 under the simple assumption that the verifying analysis has independent errors. Thus, for $\tau > 0$, we have

$$E_p(\tau) = 2E_0 + (\alpha E_0 + S)\tau \qquad (4.4)$$

For a particular model a linear empirical fit to a plot of perceived values of E_p against values of $\tau > 0$ for a day or so determines first E_0 from the intercept and then S from the slope. Greater confidence is achieved by, at the same time, fitting values of $E_p(\tau)$ for a null model, namely, those of persistence forecasts. The intercept should be the same but the null model slope is greater and determines a value S_0. The ratio S/S_0 is a dimensionless figure of merit for a model. It must be remembered in making the linear empirical fit that Eq.(4.1) includes no effects of saturation for large errors, thus that smaller errors at shorter times should be more heavily weighted.

It is assumed that the foregoing determination of E_0 is without any benefit of assimilation methods. Thus, E_0 is the error variance of an analysis which may use climate but not a model prediction as a source of a priori information. The benefits of assimilation will be examined next.

4.4 Data assimilation

The basic idea of data assimilation is to combine information from a new set of observations with the a priori information about the state of the atmosphere available from a short-range prediction valid at the new observing time. In this way information from earlier observations is carried forward, although somewhat degraded, to provide an independent source of information to be added to that newly acquired. It is straightforward to compute an error budget for the assimilation process by using Eq. (4.1) between observation times and the general principle for compositing information at observation times.

Let now τ be the fixed pseudo-time interval of the assimilation cycle, and let E_n be the error variance after data assimilation at the nth cycle. According to Eq. (4.1), prediction over the pseudo-time interval τ leads to a prediction error variance

$$E_{\tau,n} = E_n(1 + \alpha\tau) + S\tau \qquad (4.5)$$

The introduction of new observations with error variance E_0 will lead to a new value E_{n+1} according to the general principle by which information is composited, thus

$$E_{n+1}^{-1} = E_0^{-1} + E_{\tau,n}^{-1} \tag{4.6}$$

To cast the problem in dimensionless form, let $\varepsilon_n = E_n/E_0$, $\sigma = S/\alpha E_0$, and $\beta = 1 + \alpha\tau$. Then Eqs. (4.5) and (4.6) may be combined to give the iterative expression

$$\varepsilon_{n+1} = [1 + \{\beta\varepsilon_n + \sigma\alpha\tau\}^{-1}]^{-1} \tag{4.7}$$

As n increases, an equilibrium level

$$\varepsilon = \lim_{n\to\infty} \varepsilon_n = \lim_{n\to\infty} \varepsilon_{n+1} \tag{4.8}$$

is reached which is the factor by which data assimilation reduces the observational error variance E_0. It is easy to deduce from Eqs. (4.7) and (4.8) that ε must satisfy the quadratic equation

$$\varepsilon^2 + \eta(\sigma - 1)\varepsilon - \eta\sigma = 0 \tag{4.9}$$

where $\eta = \alpha\tau/\beta$ is a dimensionless parameter depending only on the assimilation time interval. The relevant root of Eq. (4.9) is given by

$$\varepsilon = [\eta\sigma + \{\eta(\sigma - 1)/2\}^2]^{1/2} - \{\eta(\sigma - 1)/2\}. \tag{4.10}$$

and is displayed in Fig. 1 as a function of σ for values of η corresponding to assimilation time intervals of 0.25 day and 0.50 day and for $\alpha = 0.5545$ day^{-1}.

Error growth results from an early GISS research model (Druyan, 1974) were fitted by Eq. (4.4) both for 500 mb height errors and velocity errors (Leith, 1978). For height errors, the resulting values are $E_0 = 200$ m^2, $S = 900$ m^2 day^{-1} and thus $\sigma = 8.25$. For velocity errors, the fitting parameters are $E_0 = 15$ m^2 sec^{-2}, $S/\alpha = 17.3$ m^2 sec^{-2} day^{-1} and thus $\sigma = 2.1$. More recently, Bengtsson (1981) reports for the ECMWF operational prediction model 5oo mb height error values of $E_0 = 150$ m^2 and $S = 400$ m^2 day^{-1} with $\sigma = 4.8$. It is not clear, however, whether E_0 in this case reflects already the benefits of assimilation.

Fig. 1 shows quantitatively how a decrease in model error sources leads to an improved equilibrium error variance. The greatest benefits appear to accrue when $\sigma = S/\alpha E_0$ is reduced to less than 1. It also appears that velocity errors may be decreased by assimilation more than are height errors.

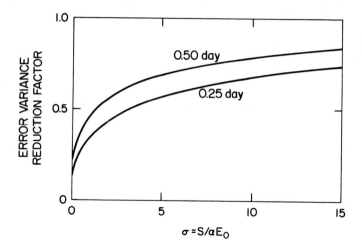

Fig. 1. Assimilation equilibrium error variance reduction ε vs. dimensionless model error source rate σ. Inherent error growth rate parameter $\alpha = 0.5545$ day^{-1}

4.5 Error growth saturation

It is a well known phenomenon that if either rms error or error variance for forecasts are plotted as a function of forecast period the curves approach asymptotically an upper level related to the climate variance. The reasons for this are clear for at late time there is vanishing correlation between the forecast anomaly $\underline{x}_f$ and the true anomaly $\underline{x}_t$. In this limit the error covariance matrix becomes

$$\underline{E} \rightarrow <(\underline{x}_f - \underline{x}_t)(\underline{x}_f - \underline{x}_t)^*>$$

$$= <\underline{x}_f \underline{x}_f^*> + <\underline{x}_t \underline{x}_t^*>$$

$$= \underline{X}_f + \underline{X}_t \qquad (4.11)$$

and any derived scalar variance will likewise give

$$E \rightarrow X_f + X_t. \qquad (4.12)$$

If the climate generated by a forecasting model when run for a long time is the same as the true climate then $X_f = X_t$ and $E \rightarrow 2X_t$ or for rms error $E^{1/2} \rightarrow 2^{1/2} X_t^{1/2}$.

Clearly in this case in which knowledge of the climate has not entered into the forecasting process the forecast error at long range reaches a value greater than that of a climate mean forecast, namely $\underset{\sim}{x}_f = \underset{\sim}{0}$, for which the error is only X_t. As described in Section 3, however, it is possible to introduce climate as an additional source of information by regression filtering and then saturation of forecast error variance occurs at a lower level with $E \to X_t$.

This large difference in saturation levels provides a warning that error variance can depend on any filtering or smoothing that has been carried out advertently or not. This suggests that the most reliable error variance for verification purposes is the irreducible value resulting from optimal regression filtering.

No saturation effects have been included in the simple error budget equations so far derived. For empirical fitting to initial error growth curves and for analysis of the assimilation process this has not been important. Perhaps the simplest way to add the effects of saturation is to add climate information or accuracy to that of the model forecast. Eq (4.2) when written in terms of the dimensionless pseudo-time $\bar{\tau} = \alpha\tau$ and the dimensionless parameter $\sigma = S/\alpha E$ becomes

$$E(\tau) = E_0[1 + (1+\sigma)\bar{\tau}] . \tag{4.13}$$

The associated model accuracy equation becomes

$$A(\tau) = A_0 [1 + (1+\sigma)\bar{\tau}]^{-1} \tag{4.14}$$

which approaches zero as $\bar{\tau} \to \infty$. Eq (4.14) may be modified for saturation effects, in the case that regression filtering is used, to give

$$A(\tau) = A_0[1 + (1+\sigma)\bar{\tau}]^{-1} + A_t \tag{4.15}$$

whence Eq (4.13) is modified to become

$$E(\tau) = \{E_0^{-1}[1 + (1+\sigma)\bar{\tau}]^{-1} + X_t^{-1}\}^{-1} \tag{4.16}$$

Note that $E(\tau) \to X_t$ as $\bar{\tau} \to \infty$.

4.6 Anomaly correlation

The correlation between forecast and observed anomalies is a frequently used measure of forecasting skill which is related to the various measures already discussed. Consider first the anomaly covariance matrix

$$\underline{C} = <\underline{x}_f \, \underline{x}_t^*> \qquad (4.17)$$

An expansion of the error covariance matrix $\underline{E}$ leads to

$$\underline{E} = <(\underline{x}_f - \underline{x}_t)(\underline{x}_f - \underline{x}_t)^*>$$

$$= \underline{X}_f + \underline{X}_t - [\underline{C} + \underline{C}^*]$$

A scalar covariance measure based solely on diagonal elements becomes then

$$C = \tfrac{1}{2}[X_f + X_t - E] \qquad (4.19)$$

or

$$C = X_t [1 - \tfrac{1}{2}(E/X_t)] \qquad (4.20)$$

when $X_f = X_t$. In this case the anomaly correlation is the bracketed term in Eq (4.20) namely

$$r = 1 - \tfrac{1}{2}(E/X_t) \qquad (4.21)$$

As $t \to \infty$, $E \to 2X_t$ and $r \to 0$.

5. Statistical sampling fluctuations

5.1 Introduction

The discussion so far has been probabilistic in nature in that first and second moments of all probability density distributions have been presumed to be known. In practice, however, we must estimate these by computing error statistics from a finite sample of forecasts. These estimates are subject to statistical sampling fluctuations that occur for finite samples even if the infinite ensemble from which they are drawn remains unchanged. These sampling errors tend to mask any small change in ensemble properties that we are trying to detect. This is a familiar problem for the analysis of climate sensitivity experiments in which one tries to detect small changes in model climate properties in response to changes in the model structures. We consider here the analogous problem for forecast sensitivity experiments. I will first summarize results from sampling theory. These are discussed in more detail in many standard texts such as that of Cramer (1945).

5.2 Statistical sampling theory

Let the random variable x have mean $<x> = 0$ and variance $<x^2> = X$. We draw a sample of n elements $(x_1, x_2, \ldots, x_n)$ from the infinite ensemble with each member equally likely to be chosen. The sample average

$$\bar{x} = \frac{1}{n} \sum_{i=1}^{n} x_i \tag{5.1}$$

should approach the ensemble mean $<x> = 0$ as n increases and the sample becomes increasingly representative. As a sum of random variables $\bar{x}$ is itself a random variable whose distribution is determined by an infinite number of repetitions of the sampling process. We can compute the first two moments of $\bar{x}$ and find

$$<\bar{x}> = \frac{1}{n} \sum_{i=1}^{n} <x_i> = 0$$

$$\bar{X} = <\bar{x}^2> = \frac{1}{n^2} \sum_{i=1}^{n} \sum_{j=1}^{n} <x_i x_j> = \frac{1}{n} X \tag{5.2}$$

Thus $\bar{x}$ has the same mean as x but its variance is less by a factor $1/n$. As $n \to \infty$ we see that $\bar{X} \to 0$, and it is in this sense that $\bar{x} \to 0$. An unbiased estimate of X is provided by the sample statistic

$$v = \frac{1}{n-1} \sum_{i=1}^{n} (x_i - \bar{x})^2 \tag{5.3}$$

since

$$<v> = \frac{1}{n-1} \sum_{i=1}^{n} <(x_i - \bar{x})^2>$$

$$= \frac{1}{n-1} \sum_{i=1}^{n} [X - \frac{2}{n}X + \frac{1}{n}X] = X \tag{5.4}$$

its second moment is

$$<v^2> = \frac{1}{(n-1)^2} \sum_{i=1}^{n} \sum_{j=1}^{n} <(x_i - \bar{x})^2 (x_j - \bar{x})^2> \tag{5.5}$$

which for a normal distribution of x can be evaluated as

$$\langle v^2 \rangle = X^2 + \frac{2}{n} X^2 \tag{5.6}$$

The variance of v is then given by

$$\langle v^2 \rangle - \langle v \rangle^2 = \frac{2}{n} X^2 \tag{5.7}$$

which also tends to zero as $n \to \infty$.

In summary although $\bar{x}$ of Eq (5.1) and v of Eq (5.3) are unbiased estimates of $\langle x \rangle$ and X they have associated sampling error variances of $\frac{1}{n} X$ and $\frac{2}{n} X^2$ respectively. The corresponding rms sampling errors are, of course, the square root of these, and if an observed change in $\bar{x}$ or v is small in magnitude compared to these, we may doubt its statistical significance as indicating a change in $\langle x \rangle$ or X.

In application to forecast sensitivity experiments $\langle x \rangle$ becomes the forecast bias $\langle x_f \rangle$ and X the error variance E. We have for convenience assumed that $\langle x_f \rangle = 0$, but in fact the climate mean of a model usually differs somewhat from that of the atmosphere and thus of the initial states. The model forecasts tend to drift toward the model climate mean in a few days introducing a climate mean drift bias. Experiments on model improvements designed to decrease this bias must be interpreted in the light of sampling errors of the mean which have a variance E/n.

We are of course, also interested in changes that will reduce E as estimated by v, but here we must face sampling errors of the variance with, in turn, a variance of $(2/n) E^2$ or standard error of $(2/n)^{1/2} E$. The significance of any change δE will depend on the dimensionless ratio

$$w = (n/2)^{1/2} (\delta E/E) \tag{5.8}$$

Eq (5.8) may be combined with Eq (4.16) to determine the best forecast range for determining, say, the effect of a change in the model error source rate S. If we let

$$\epsilon_c = X_t / E_0 \tag{5.9}$$

and

$$\bar{\epsilon} = E/E_0 \tag{5.10}$$

then Eq (4.16) may be written in dimensionless form as

$$\overline{\epsilon}(\tau) = \{ 1 + (1+\sigma)\overline{\tau}]^{-1} + \epsilon_c^{-1}\}^{-1} \qquad (5.11)$$

According to Eq (5.8) the detectability of a change in σ is proportional to

$$\kappa(\tau) = \frac{1}{\overline{\epsilon}} \frac{\partial \overline{\epsilon}}{\partial \sigma} = \overline{\epsilon} \lambda^{-2} \overline{\tau} \qquad (5.12)$$

where $\lambda = 1 + (1+\sigma)\overline{\tau}$.

The quantity κ approaches 0 as $\overline{\tau} \to 0$ where the effect of model errors has not yet been felt and as $\overline{\tau} \to \infty$ where climate variance dominates the error variance. In between it has a maximum at $\overline{\tau} = \overline{\tau}_m$ where

$$\overline{\tau}_m = \frac{(1+\epsilon_c)^{1/2}}{1+\sigma} \qquad (5.13)$$

Evaluation of Eq. (5.13) for typical values such as $\sigma = 6$ and $\epsilon_c = 48$ gives a value $\overline{\tau}_m = 1$ and an optimal forecast time interval $t_m = 1.25$ days.

The anomaly correlation measure r defined by Eq. (1.21) is also subject to sampling errors. The correlation coefficient r is bounded with $-1 \leq r \leq 1$, and its sampling probability distribution is a quite complicated function of the ensemble correlation ρ, the sample correlation r, and n. This distribution is considerably simplified by the transformation of variables introduced by Fisher (1941)

$$\zeta = \frac{1}{2} \log \frac{1+\rho}{1-\rho} \qquad (5.9)$$

$$z = \frac{1}{2} \log \frac{1+r}{1-r} \qquad (5.10)$$

in terms of which the distribution of z is more nearly normal with moments

$$<z> = \zeta + \frac{\rho}{2(n-1)} \qquad (5.11)$$

$$Z = <z^2> - <z>^2 = \frac{1}{n-3} \qquad (5.12)$$

to lowest order in $1/n$.

5.3 Effective sample size

Sampling theory is based on knowledge of the number n of independently drawn members of the infinite ensemble. A special problem arises in applications to atmospheric statistics where time and space correlations can decrease the effective sample size. For climate studies based on time averages over a time interval T this has led to the definition of an effectively independent sampling time T_0 given (Leith, 1973) as

$$T_0 = \int_{-\infty}^{\infty} R(\tau) d\tau$$

where $R(\tau)$ is a characteristic time-lagged correlation function. For many atmospheric variables T_0 is of the order of a week. The effective sample size is then given asymptotically for large T as $n = T/T_0$. This analysis was appropriate for estimation of a mean. For estimation of a variance a better time is

$$T_0' = \int_{-\infty}^{\infty} R^2(\tau) d\tau \approx T_0/2$$

The number n is greater by about a factor of 2 for variance than for mean estimates which tends to cancel the factor 2 appearing in Eq (5.7).

Similar arguments are appropriate for spatial correlations when statistics are generated by averaging over space. By analogy one defines an effectively independent averaging area as

$$A_0 = \int_{-\infty}^{\infty} \int_{-\infty}^{\infty} R(x,y) dx\, dy$$

for first moments and

$$A_0' = \int_{-\infty}^{\infty} \int_{-\infty}^{\infty} R^2(x,y) dx\, dy \approx A_0/2$$

for second moments. The approximations for T_0' and A_0' are based on observations that time-lagged correlations are nearly decaying exponentials while space correlations are nearly Gaussian.

6. Conclusion

These lecture notes have been primarily concerned with the more theoretical aspects of the subject and have ignored many practical difficulties in forecast verification. I hope that the following key ideas can serve as a guide for future work.

1) With appropriate definitions error and accuracy can be treated as additive (Section 3)

2) Efforts to construct error budget equations can sharpen our understanding of the total forecasting system (Section 4)

3) Our knowledge of the state of the atmosphere at any time depends to a considerable extent on the forecasting model used to bring forward past information (Section 4.4).

4) We must not be misled by statistical sampling fluctuation when testing possible improvements in a forecasting technique (Section 5)

References

Bengtsson, L. 1981 Medium range weather forecasting at ECMWF and remaining problems. Extended abstract, Symposium on Current Problems of Weather Prediction, Vienna, June, 1981.

Cramer, H. 1945. Mathematical Methods of Statistics. Princeton University, Press, chaps 27-29.

Druyan, L.M. 1974. Short-range forecasts with the GISS model of the global atmosphere. Mon. Weather Rev., 102, 269-279.

Fisher, R.A. 1941 Statistical Methods for Research Workers. Eigth ed., Edinburgh and London

Gauss, K.F. 1809 Theory of the Motion of the Heavenly Bodies Moving about the Sun in Conic Sections. C.H. Davis Transl., republished by Dover Publ., 1963.

Leith, C.E. 1978 Objective methods for weather prediction. Ann. Rev. Fluid Mech., 10, 107-128.

Leith, C.E. 1973 The standard error of time-averaged estimates of climatic means. J. Appl. Meteor., 12, 1066-1069.

Leith, C.E. 1975 Statistical-dynamical forecasting methods. In Seminars on Scientific Foundation of Medium Range Weather Forecasts, Reading, ECMWF.

Bifurcation Mechanisms and Atmospheric Blocking

E. KÄLLÉN*

Abstract

The bifurcation properties of low order, barotropic models with orographic and a Newtonian type of vorticity forcing are reviewed. The low order model results that multiple equilibria develop as a result of a sufficiently strong wave, orographic forcing and that a suitably positioned wave vorticity forcing can enhance this bifurcation mechanism are verified with a high resolution, spectral model. The high resolution model is integrated in time to find stable steady-states. Bifurcations into multiple equilibria appear as sudden jumps in the amplitudes of the model components, when the forcing is slowly changing in time.

Diagnostic studies of the mountain torque and the eddy activity in the Northern Hemisphere during winter are compared with the occurrence of blocked flows to support the blocking mechanism originally proposed by Charney and DeVore (1979).

1. INTRODUCTION

In recent years there has been a renewed interest in the study of low-order systems to gain some insight into nonlinear mechanisms present in the atmosphere. The basic procedure used when studying a low-order system is to expand the space dependent quantities into a series of orthogonal functions and to truncate this expansion by just retaining a few components. Each component is thought of as representing a certain scale of motion and, by inserting the truncated expansion into an equation of motion, one can study the nonlinear interactions between the scales considered. One thus neglects

---

*Present affiliation: Department of Meteorology, University of Stockholm, Sweden

all interactions with spectral components not taken into account. This is of course a serious limitation of low-order systems, but it is nevertheless believed that a study of such systems is one way of getting an insight into the nonlinear mechanisms present in the atmosphere. The models discussed here only take wave-mean flow interactions into account and this approximation is discussed by Herring (1963).

When studying a bifurcation problem with a forced, dissipative model the low order approximation can also be justified as follows. When the forcing is weak the response is small and the linear dissipative and dispersive terms dominate over the nonlinear terms which are of a higher order and the model may be regarded as quasi-linear. For a stronger forcing the response is increased and the nonlinear terms, which in the models studied here are quadratic, become increasingly important. In spectral space, a weak forcing of a certain wavenumber gives a strongly peaked spectrum, while a stronger forcing distributes energy over a wider range of spectral components. At the forcing value where the first bifurcation occurs it is thus the spectral components neighbouring the forced one which are likely to dominate the nonlinearity. It may then be possible to investigate this first bifurcation by only taking components into account which are close in spectral space. To verify results found with a low-order model one should also perform experiments with a high-resolution, fully nonlinear model. These experiments must, however, be guided by the qualitative properties of the low-order system.

To study the effect of orographic forcing on the nonlinear energy transfer between the larger scales of motion, Charney and DeVore (1979) (hereafter called CdV) extended Lorenz's (1960) barotropic β-plane model to include orographic forcing. With a low-order system they showed that for a given forcing it was possible for the flow to arrange itself in several equilibrium states, some stable and some unstable. The multiplicity of equilibrium states is associated with the resonance occurring when the Rossby wave, generated by the zonal flow over the orography, becomes stationary. Through the nonlinear interaction between the zonal flow and the wave components of the flow and due to the orography, the components can arrange themselves in two stable equilibria, one close to resonance with a large amplitude wave and a weak zonal flow, the other with a strong zonal flow and a weaker wave component. The large amplitude wave flow may be associated with a blocked flow in the atmosphere.

This result was first derived for a β-plane, channel model with reflecting side walls, but later studies by Davey (1980, 1981) and Källén (1981) have shown that the same type of mechanism can be found with an annular or a spherical geometry. Trevisan and Buzzi (1980) also showed the same phenomenon using a different expansion method on a β-plane geometry.

The basic equation used for the models of all these studies is the quasi-geostrophic, barotropic vorticity equation with linear dissipation, a Newtonian type of vorticity forcing and orographic forcing. In a non-dimensional form this equation is

$$\frac{\partial \zeta}{\partial t} - J(\zeta + f + h, \psi) + \varepsilon(\zeta_E - \zeta) \tag{1}$$

where ζ is the non-dimensional vorticity, f is the planetary vorticity, ζ_E the vorticity forcing, ψ the stream function and h is a parameter related to the orographic effects. The non-dimensional time is given by t and ε is the dissipation rate. The orographic forcing may be introduced as a forced vertical velocity of the lower boundary in an equivalent barotropic model, in which case h is related to a dimensional mountain height m via

$$h = C \cdot \frac{m}{H} \tag{2}$$

where H is the scale height of the equivalent barotropic atmosphere and C is a constant which depends on the equivalent barotropic assumption. For normal atmospheric conditions the constant C is approximately equal to one. For details of the derivation of Eq.(1) see Källén (1981). To describe the phase speeds of the long waves properly, a correction term in the time derivative dependent on the Rossby radius of deformation should be included in Eq.(1), (see Wiin-Nielsen, 1959). Since the model is not intended to simulate this aspect of the long waves with any accuracy the correction term will be omitted and the equation will be simpler to handle. The correction will only affect the phase speeds of moving long waves and not the stationary states which this study mostly deals with.

The nonlinearity of the model is contained in the term involving the Jacobian $(J(\zeta + f + h, \psi))$, and one way of investigating the nonlinear properties of this model is to expand the space dependent variables in a series of orthogonal functions, $F_\gamma(\underline{x})$, where $\underline{x}$ is the space vector. It is convenient to choose the F_γ's to be eigenfunctions of the Laplacian operator, because $\zeta = \nabla^2 \psi$. The exact functional form of the F_γ's of course depends on the geometry of the model and the boundary conditions.

The variables to be expanded are the vorticity (also gives the streamfunction), vorticity forcing and the orography. The expansion may be written

$$\begin{pmatrix} \zeta \\ \zeta_E \\ h \end{pmatrix} = \sum_\gamma \begin{pmatrix} \zeta_\gamma(t) \\ \zeta_{E,\gamma} \\ h_\gamma \end{pmatrix} F_\gamma(\underline{x}) \qquad (3)$$

For a β-plane, channel model trigonometric functions can be used and

$F_{m,n}(x,y) = e^{i(mx+ny)}$, x and y being the Cartesian coordinates (CdV). In an annular geometry, Bessel functions are involved (see Davey, 1981) and on the sphere the F_γ's may be written

$$F_{m,n}(\mu,\lambda) = P_n^m(\mu) \, e^{im\lambda} \tag{4}$$

where μ is the sine of latitude (ϕ), λ is the longitude and $P_n^m(\mu)$ are associated Legendre functions. Most of the results discussed in this paper will refer to a spherical geometry as in Källén (1981) and thus the expansion functions given by Eq.(4) are the appropriate ones.

A low-order model may be formulated by inserting the expansion Eq.(3) in the model Eq.(1) and truncating the expansion at a very low order just leaving a few components to describe the fluid motion. Each component may be thought of as describing a certain scale of motion and only the nonlinear interactions between the scales involved in the low order system are taken into account.

To study the effects of the orography on the interaction between the waves and the mean zonal flow, at least one purely zonal component and two wave components have to be included. The mathematical structure of such a minimal system is independent of the geometry and multiple equilibrium states may be found even in such a simple model, as first pointed out by CdV. When additional components are included the geometry will affect the structure of the equations, but the basic mechanism for the formation of multiple equilibrium states still remains. In the following section a minimal system will be analyzed, following the basic idea of CdV. Section 3 will deal with a combination of orographic and direct wave vorticity forcing as discussed by Källén (1981). The direct wave vorticity forcing is thought to represent the time mean effect of the small scale baroclinic eddies on the long waves in the atmosphere. The effect of baroclinicity is thus taken into account only parametrically. The barotropic nature of the models prevents us from taking any baroclinic effects into account directly and the wave vorticity forcing is only a crude first order approximation to the effects of baroclinic waves. Calculations from observations (Lau, 1979) do however show that the approximation is reasonable. Some investigations have been made on how baroclinic effects may modify the multiple equilibria obtained with barotropic models (Charney and Strauss, 1980 and Roads, 1980) and recently a study of the interaction between transient baroclinic waves and orographically forced stationary waves has appeared (Reinhold, 1981). In Section 4 a verification of some of the drastic assumptions made when dealing with low-order models will be made and in Section 5 we will try to relate the bifurcation properties of the simplified models to diagnostic studies of the atmosphere in connection with blocking events.

2. WAVE-MEAN FLOW INTERACTIONS VIA THE OROGRAPHY

The simplest possible low-order model in which one may investigate the nonlinear coupling between the zonal mean flow and the eddies through the effect of the orography, is a model which involves one component describing a purely zonal flow and two components describing the phase and amplitude of a wave. On the sphere the streamfunction may thus be written

$$\psi = -u_o(t) P_1(\mu) + (x_1(t) \cos\ell\lambda + y_1(t) \sin\ell\lambda) P_n^\ell(\mu)$$

where u_o, x_1 and y_1 are the time dependent amplitudes of each flow component. To drive the zonal flow, a vorticity forcing is introduced in the $P_1(\mu)$ component with an amplitude, u_{oE}. The orography is assumed to be present with an amplitude, h_1, in the $\cos \ell\lambda P_n^\ell(\mu)$ component. Inserting these expansions of the stream function, vorticity forcing and the orography into the barotropic vorticity equation (1), a set of three ordinary differential equations governing the time evolution of the model is obtained,

$$\dot{u}_o = h_1 \delta_1 y_1 + \varepsilon(u_{OE} - u_o)$$
$$\dot{x}_1 = (\beta - \alpha u_o)y_1 - \varepsilon x_1 \qquad (5)$$
$$\dot{y}_1 = -h_1 \delta_2 u_o - (\beta - \alpha u_o)x_1 - \varepsilon y_1$$

The coefficients appearing in (5) are defined

$$\alpha = \sqrt{3}\ell\,[1-2/n(n+1)] \;,\; \beta = \frac{2\ell}{n(n+1)} \;,\; \delta_1 = \frac{\sqrt{3}}{4}\ell,\; \delta_2 = \frac{\sqrt{3}\ell}{n(n+1)} \;.$$

These expressions for the coefficients have been derived for a spherical geometry, but in fact the structure of the equations is exactly the same for a β-plane or an annular geometry, the only difference between the different geometries lie in the expressions for and values of the coefficients. The results shown in this section, may thus equally well be applied to the other types of geometry.

To investigate the mathematical properties of the nonlinear system of equations (5), we first determine the steady-state values of u_o, x_1 and y_1. Setting the time derivatives equal to zero and solving the resulting system of equations for one of the steady-state amplitudes (u_o), we arrive at the following equation in $\bar{u}_o$ (hereafter an overbar will denote a steady-state amplitude)

$$\bar{u}_o^3 - \bar{u}_o^2(u_{OE} + \frac{2\beta}{\alpha}) + \frac{\bar{u}_o}{\alpha^2}[\delta_1\delta_2 h_1^2 + \varepsilon^2 + \beta^2 + 2\beta u_{OE}] = u_{OE}\frac{(\varepsilon^2 + \beta^2)}{\alpha^2} \qquad (6)$$

which also may be written

$$u_{OE} = \frac{\delta_1\delta_2 h_1^2 \bar{u}_o}{\varepsilon^2 + (\beta - \alpha\bar{u}_o)^2} + \bar{u}_o \qquad (7)$$

The polynomial form (6) of the steady-state equation shows that we can at the most have three steady-states for certain values of the forcing parameters. The other form of the steady-state equation, (7), allows us to investigate by graphical methods how the number of steady-states varies with the forcing parameters. Note that on the left hand side of (7), u_{OE} is a velocity. This velocity is the purely linear response of the u_o-component when there is no orography.

The example chosen to illustrate Eq. (7) is one in which ℓ =3 and n=4. This implies an orography with a zonal wavenumber three and a maximum in mid-latitudes. Fig. 1 is a plot of the curves given by Eq.(7) for some values of h_1. Both axes are given in non-dimensional and dimensional units. The dimensional units are m/s and correspond to the windspeed of the u_o component at 45° latitude.

The curve in Fig. 1 for h_1 = 0 is a straight line and for increasing values of h_1 we obtain a family of curves, some having a section with a negative slope. As each point on the curves represents a steady-state of the system given by Eqs. (5), a negatively sloping section of a curve implies that within a certain forcing parameter (u_{oE}) interval it is possible to have three steady-states. Taking as an example the curve for h_1 = 0.20 it can be seen from Fig. 1 that in the interval 0.135 < u_{oE} < 0.158 three steady-states exist. The stability properties of each steady-state are found by linearising Eq. (5) around each steady-state, and finding the eigenvalues of the linearized equations. An investigation of this kind gives stability properties as indicated by full (stable) and dashed (unstable) lines in Fig. 1. In Fig. 2 it may also be seen that the bifurcation from one to three steady-states occurs for a value of h_1 somewhere between 0.10 and 0.15.

The possibility of having two stable equilibrium states for constant values of the forcing parameters is due to the nonlinear coupling between the zonal flow and the waves. The coupling by itself, however, without the effect of the orography, does not give rise to multiple steady-states in Eq.(5). It is the influence of the orography which is crucial in creating an instability which gives the possibility of having multiple steady-states. CdV called

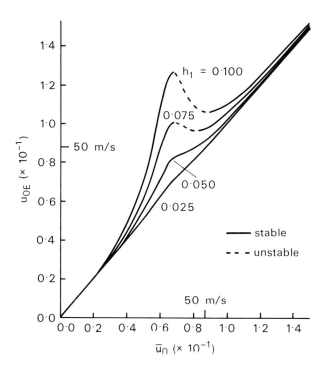

Fig. 1. Steady-state curves for the low order model of section 2 with different values of the orographic parameter. The horizontal axis gives the amplitude of the $\bar{u}_o$-component, both in non-dimensional and dimensional units. On the vertical axis the forcing is also given in both units. The dimensional units are the flow velocity at 45° latitude. Each curve corresponds to a certain value of the orographic parameter, h_1, and a numerical evaluation of the eigenvalues shows stability properties as indicated by full (stable) and dashed (unstable) lines. For further explanations, see text.
Parameter values: $\varepsilon = 0.06$, $\ell = 3$, $n_1 = 4$

this a form drag instability, where the form drag refers to the effect of the orography in this simple model.

One of the stable steady-states is close to a resonant flow configuration, and in this steady-state the wave component has a high amplitude. The phase of the wave is such that there is a high drag across the orographic ridges and thus energy is transferred from the zonal forcing, via the effect of the orography, to the wave components of the flow. In Fig. 1 this type of a steady-state falls on the left hand part of the diagram where the response in the zonal component (u_o) is much less than the forcing.

The steady-states on the right hand side of the unstable region have a more intense zonal flow and a less marked wave component. In these steady-states the orographic drag is much lower, both due to the lower amplitude of the wave component and a different phase of the wave.

The steady-state with a high wave amplitude may be associated with a blocked flow in the atmosphere. The wave ridge occurs downstream of the orographic ridge, and the persistence of blocking ridges in the atmosphere may be due to the stable wave-mean flow interaction described by this simple model. The model also predicts that there may be another stable flow configuration without a high amplitude wave, but a much more pronounced zonal flow. Which one of these steady-states the flow settles into is crucially dependent on the initial state of the flow. CdV offered this feature as a possible explanation for the observed variability in the frequency of the occurrence of blocked flows.

3. A COMBINATION OF WAVE VORTICITY AND OROGRAPHIC FORCING

In the model described in the previous section, the waves of the flow are only due to the interaction between the zonal flow and the orography. In the atmosphere there are numerous other processes which generate waves and in midlatitudes the most important one is the baroclinic instability process. Baroclinic waves have a characteristic wavelength which is shorter than the scales involved in the low order models of this study, but seen as an effect on the time mean flow the energy generated by baroclinically unstable waves on the shorter scales is transported in the spectrum through nonlinear processes and thus also exports kinetic energy to the longer waves (Saltzmann, 1970 and Steinberg et al, 1971). In a barotropic model it is impossible to describe these baroclinic effects explicitly, but as a first approximation one may be able to take the long wave effect into account by introducing a wave vorticity forcing in the same components as the orography. The wave vorticity forcing should thus be seen as the time mean effect of cyclone waves rather than a direct diabatic heating.

A wave vorticity forcing can easily be introduced into the low order model of the previous section by adding εx_{1E} and εy_{1E} to the right hand sides of the equations for $\dot{x}_1$ and $\dot{y}_1$ of Eq.(5). The steady-states may be analyzed in the same way by writing the steady-state equation in terms of the zonal forcing (u_{oE}) as a function of the zonal response (u_o) with the orography (h_1), the amplitude ($\sqrt{x_{1E}^2 + y_{1E}^2}$) and the phase ($\tan^{-1}\frac{y_{1E}}{x_{1E}}$) of the wave vorticity forcing as parameters. In Källén (1981) this was done, but with a sightly more complicated model. Two extra wavecomponents and one extra zonal component was included to take into account some of the interactions with unforced parts of the spectrum. It is still possible to solve for the steady-states analytically in such a model according to the procedure given in Källén (1981). We will not go into any detail here regarding the solution method, but only display some of the results.

The streamfunction expression used in the examples is

$$\psi(\mu, \lambda, t) = -u_o(t) P_1(\mu) + z(t) P_3(\mu)$$

$$+ (x_1(t) \cos 3\lambda + y_1(t) \sin 3\lambda) P_4^3(\mu) +$$

$$+ (x_2(t) \cos 3\lambda + y_2(t) \sin 3\lambda) P_6^3(\mu)$$

and the vorticity forcing is given by

$$\psi_E(\mu, \lambda) = -u_{oE} P_1(\mu) +$$

$$+ (x_{1E} \cos 3\lambda + y_{1E} \sin 3\lambda) P_4^3(\mu) \; .$$

The orography is the same as in the previous section,

$$h = h_1 \cos 3\lambda \, P_4^3(\mu)$$

The low order system is thus made up of six ordinary differential equations, which govern the time evolution of the model. Because of the choice of components, no direct wave-wave to wave interactions are allowed but energy transfer from one wave-component to the other may take place via the zonal flow. The zonal component with the amplitude z describes a sheared zonal flow and it is via this zonal component that energy may be transferred through flow-flow interactions from the forced to the unforced wave components.

A plot of the steady-states of the system is given in Fig. 2. On the horizontal axis the steady-state response is given in terms of the amplitude of the sheared zonal component, z. The vertical axis gives the zonal momentum forcing, u_{oE}.

The orographic forcing is fixed at a value of 0.05 while the wave vorticity forcing has a varying amplitude, but the phase in relation to the orography is fixed. Multiple steady-states of the system are identified in the figure by the condition that a horizontal line should have multiple intersections with one of the steady-state curves. The forcing parameters are in a range where multiple steady states are just possible, i.e. close to the first bifurcation point. For smaller values of the forcing parameters the nonlinear system behaves quasi-linearly, with just one steady state for a certain value of the forcing parameters. By linearizing the system around a steady-state and computing the eigenvalues of the matrix governing the linearized motion around each of the steady-states, stability properties are found as indicated in Fig. 2. It may be noted that no Hopf-bifurcations (see Marsden and McCracken, 1979) indicating the existence of limit-cycles around the steady-states have been found in this low order model for reasonable values of the forcing parameters.

For the curve marked I in Fig. 2 the wave vorticity forcing is set to zero and the only forcing of the model is in the orography and the zonal momentum. For the orography height chosen in Fig. 2 there is only one, stable steady-state for all values of the zonal momentum forcing. For higher values of the orographic parameter multiple steady-states are possible within certain ranges of values of the zonal momentum forcing. For further details of this, see Källén (1981). Another way of obtaining a region of multiple equilibria is to include vorticity forcing in one of the wave components and this is shown with steady state curve II in Fig.2. The phase of the wave vorticity forcing is -90°, i.e. positive (cyclonic) vorticity forcing on the

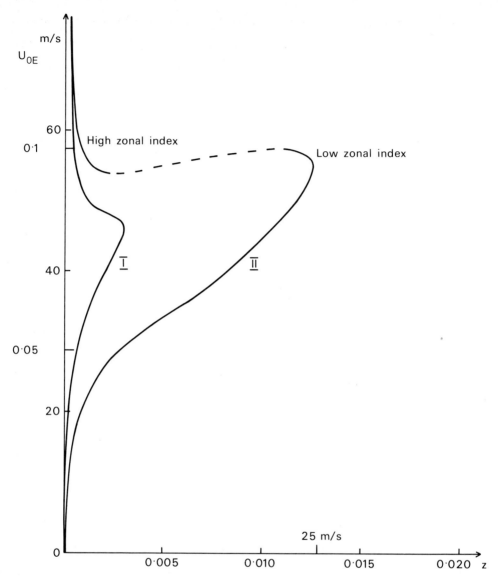

Fig. 2. Steady-state curves for the low order model of section 3. The abscissa gives the response in terms of the amplitude of one of the zonal components (z), the ordinate gives the strength of the zonal forcing (u_{OE}). Both axes are given in dimensional and non-dimensional units, the dimensional units being taken as the zonal average at 45° latitude. The height of the orography, $h_1 = 0.05$ and the dissipation rate, $\varepsilon = 0.06$. For curve I there is no wave vorticity forcing while for curve II there is a wave vorticity forcing in the same component as the orography ($\ell=3$, $n_1=4$). The amplitude of the wave forcing of curve II is $x_{1E}= 0.$, $y_{1E}=-0.015$ which corresponds to 20 m/s in terms of a zonally averaged absolute value of the meridional wind at 45° latitude and a phase angle of -90°. Stability properties as indicated by dashed (unstable) and full (stable) lines

leeward side of the highs in the orography. In Källén 1981 it was shown that this phase angle is the most favourable one for bifurcations to occur. The characteristics of the steady-states within the region of multiple equilibria can be found in the example displayed in Fig. 3. The steady-states on the left hand, upper branch of curve II in Fig. 2 have a marked zonal flow and a rather weak wave component as the top flow of Fig. 3. The steady-states on the right-hand branch, on the other hand, have a much stronger wave component and a weaker zonal flow as the bottom flow of Fig. 3. These latter steady-states can be associated with low index atmospheric circulations, i.e. blocking periods, while the steady-states on the left hand branch have the characteristics of a high index, zonal type of circulation. Examining the energetics of these two solution types it was found in Källén (1981) that in the zonal steady-state the orographic influence on the flow was much weaker than in the blocked state. In the blocked state the orography acted to transform zonal kinetic energy into wave kinetic energy in a much more intense way than in the zonal state. Furthermore, the efficiency of the flow in picking up energy from the wave vorticity forcing was markedly different. In the blocked flow the phase difference between the forced wave and the wave forcing is very small, thus giving a high amplitude response. In the zonal steady state the trough on the leeward side of the orography is further downstream from the orographic high than in the blocked case, and the response amplitude is thus lower. The unstable steady-states on the dashed part of curve II have properties somewhere intermediate between the two stable branch steady-states. The unstable steady-states are, of course, not very interesting. Because of their instability the flow will never settle into them.

It has thus been demonstrated that a combination of orographic and wave vorticity forcing can give rise to multiple equilibrium states, even when the separate effects of the two types of forcing do not show this type of

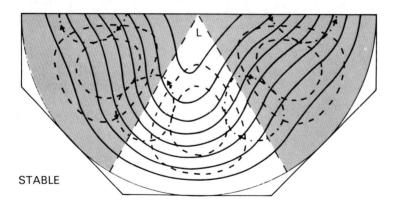

STABLE

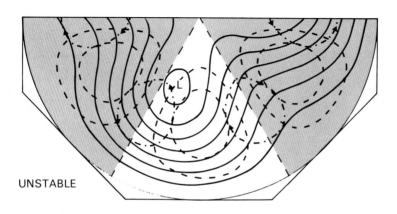

UNSTABLE

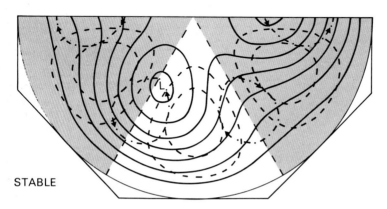

STABLE

Fig. 3. Examples of stream function fields for three steady-states within the region of possible multiple equilibria of Fig. 2 (curve II). Full lines are isolines of the streamfunction while the dashed lines are isolines for the orography. Over the hatched area the orography is above its mean value ("land areas") while otherwise it is below its mean value ("ocean area"). Dash-dotted curves with arrows showing direction of circulation indicate regions with maximum cyclonic and anti-cyclonic wave vorticity forcing

behaviour. With wave vorticity forcing alone this model does not give rise to more than one equilibrium state. The orography can be seen to act as a triggering mechanism, directing the basically vorticity forced flow into one or the other of the stable equilibria.

4. HIGH RESOLUTION EXPERIMENTS

A serious shortcoming of a severely truncated low order system is of course the lack of interaction between waves of all scales. Only a few scales of motion are taken into account and the interactions with other scales are either neglected completely or included via a bulk momentum forcing. To investigate whether the bifurcation mechanism found in a low order model is sensitive to the number of waves present in the model, the results should in some way be verified with a high resolution model. CdV showed that the multiple equilibria in their β-plane channel model could also be found in a model with an increased resolution, while (Davey 1981) pointed out that it is possible to find the multiple equilibria in a high resolution model but they do not obtain as easily as in a low order model.

To verify the results of Källén (1981) for a spherical geometry, experiments have been performed with a high resolution (horizontally) quasi-geostrophic, spectral, barotropic model. These experiments will be described here following Källén (1982).

The high resolution model was originally developed at the University of Reading, UK and it is a barotropic and quasi-geostrophically balanced version of the model described in Hoskins and Simmons (1975). The governing equation of the model is the same as in Section 2, i.e. Eq.(1). Forcing is introduced in exactly the same components as in the low-order model of Section 3 and the model is integrated in time to find the steady-state(s). For reasons of economy, most of the experiments were done with a T21 truncation ($k \leq 21$ and $n \leq 21$ in the Legendre ($P_n^k(\mu)$) representation).

However, some integrations done with a T42 truncation showed no significant differences to the T21 experiments.

To find multiple stable steady-states the time integrations are set up in the following manner. Initially the forcing is held constant for a time period of twenty days. The model, starting from a state of rest, is allowed this time to find a steady-state. After the model has settled into a steady-state the zonal momentum forcing is slowly increased or decreased in time. The time scale of this slow increase or decrease is chosen to be significantly slower than the dissipation time scale of the model (the forcing is doubled or halved in 100 days while the dissipation time scale is around 2.5 days). With this rather large dissipation rate the model stays reasonably close to a stable steady-state all of the time. If, however, a bifurcation of the type pictured in Fig. 2 occurs, the model has to make a sudden jump from one stable branch of the steady-state curve to the other when a critical value in the zonal forcing is passed. In a time plot of some of the spectral coefficients this will show up as a sudden change of the amplitude and perhaps some damped oscillations when the model settles into a steady-state on the other branch.

Results of experiments of this type are displayed in Fig. 4. The forcing parameters have values which are close to the ones used in the low order example of Fig. 2. The phase difference between the wave vorticity forcing and the orography is exactly the same, -90°. It was, however, found that in order to obtain multiple equilibrium states with a high resolution model the vorticity forcing has to be higher. This is presumably due to the fact that the existence of more components forces the energy introduced at certain wave components to spread out to all parts of the spectrum. The long waves, therefore need more energy input to reach the critical amplitudes necessary for bifurcations to occur. The orography, on the other hand, had to be

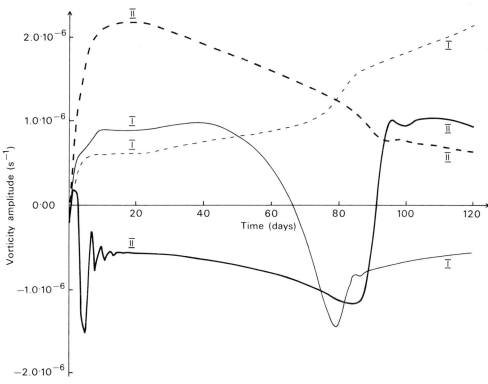

Fig. 4. Vorticity amplitudes of one wave component (k=3, n=4; solid curves) and one zonal component (k=0, n=1; dashed curves) as a function of time in the high resolution model. Curves marked I refer to the experiment with increasing zonal forcing, the ones marked II to a decreasing zonal forcing.
(c = 0.06, h = 0.025, x_{1E} = 0., y_{1E} = -0.02,

$u_{OE} = \begin{matrix} 0.05 + \\ 0.15 - \end{matrix}$ (time-20)*0.001, time in days)

lowered to avoid the formation of multiple equilibria when only zonal momentum forcing is present. The high resolution model thus appears to be more sensitive to the height of the orography than the low order model.

In Fig. 4 the amplitudes of the forced zonal and wave components are shown as functions of time for two different experiments. In one experiment (curves marked I) the zonal forcing starts at a fairly low value, well below the bifurcation "knee" of the curve in Fig.2. After day 20 the zonal forcing is increased slowly and from the amplitudes of the forced zonal component and the forced wave component it can be seen that a sudden jump occurs around day 80. The curves marked II show a similar experiment, but this time the zonal forcing is decreasing as a function of time. In these curves the jump occurs around day 90 and it should be noted that the jump does not occur at the same value of the zonal forcing as for the jump with an increasing zonal forcing. This strongly suggests that there is a certain interval in the values of the zonal forcing where multiple stable steady-states are possible. To find these steady-states the same type of integrations have been performed, but instead of increasing/decreasing the zonal forcing until the end of the integration the zonal forcing was held constant at a certain level after an increase/decrease from an initially low/high value. The final level of the forcing was chosen to be in the range of possible multiple equilibria. In Fig. 5 the resulting streamfunctions from such an experiment are displayed. The top figure is the steady-state which is reached from an initially high value of the zonal forcing, the lower part is the steady-state reached from an initially low value of the zonal forcing. Comparing these figures with the low order results of Fig. 3 it may be concluded that the streampatterns are qualitatively similar. One has a pronounced wave component and can be interpreted as a "blocked" or high amplitude wave steady-state. The other has a much more marked zonal flow and a weaker wave component. The feature displayed in Fig. 2 and shown in Källén (1981) that when orography alone is not sufficient to produce a forcing regime with multiple equilibria, a

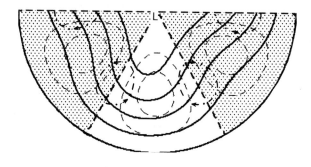

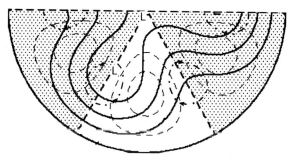

Fig. 5. Two stable streamfunction patterns having the same values of all the forcing parameters ($\varepsilon = 0.06$, $h = 0.025$, $x_{1E} = 0.0$, $y_{1E} = -0.02$, $u_{0E} = 0.095$). Full lines are isolines for the streamfunction with the same isoline interval in the two plots. Dashed lines are isolines for the orography, areas with the orography above its mean value (land areas) are hatched. The dashed-dotted curves indicate the positions and directions of maximum and minimum wave vorticity forcing

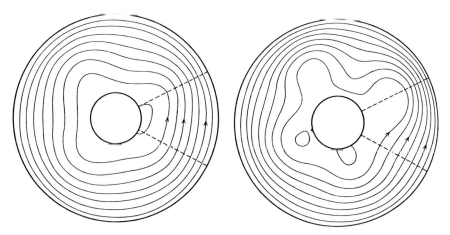

Fig. 6. Alternative stable states in an annular model with an isolated orographic ridge, taken from Davey (1981). Orographic ridge in the right hand sector between the two dashed lines. Full lines are isolines of the streamfunction

combination of orographic and wave vorticity forcing does give this possibility, is reproduced by the high resolution model. The addition of wave vorticity forcing is thus not just a linear addition of wave energy to the flow, instead it actively takes part in the nonlinear formation of multiple equilibria.

The numerical experiments with the high resolution spectral model thus strongly support the results derived from the low order model of Kallen (1981). There are, however some aspects of the low order model behaviour which are not verified by the high resolution experiments. One such behaviour is the bifurcation obtained in a low order model in the absence of orography. With a five component low order model it is possible to find multiple equilibria with only vorticity forcing (Wiin-Nielsen, 1979) on the longer waves. Experiments with the high resolution model have not shown this feature, even for very large values of the forcing parameters. The model behaves perfectly linear when only momentum forcing is applied, the response to the forcing being purely in the forced components. For shorter waves this is no longer true. Hoskins (1973) has demonstrated that for zonal wavenumbers larger than five a nonlinear instability develops which is mainly due to wave-wave interactions. For the longer waves the beta-effect acts as a stabilizing factor which prevents this type of nonlinear instability. With orography present it thus seems that a new type of instability develops as first pointed out by CdV. An intuitive reasoning which points to a possible reason for this property of the orography can be given as follows. The governing equation of the model at a steady-state may be written

$$J(\zeta,\psi) + J(h,\psi) - 2\frac{\partial \psi}{\partial \lambda} + \epsilon(\zeta_E - \zeta) = 0 \qquad (8)$$

If $h=0$ (no orographic forcing) and $\zeta_E \neq 0$ it is possible to have a steady-state where the response is in the same component as the forcing and the nonlinear term $J(\zeta,\psi)$ is zero. As mentioned above, numerical experiments with reasonable values of the vorticity forcing on the longer waves have shown that such a steady-state is stable. Equation (8) is in this case linear. This also holds if the wave vorticity forcing is applied at several low wave numbers simultaneously. If an orographic forcing is introduced ($h \neq 0$) the term $J(h,\psi)$ forces energy introduced via the vorticity forcing ζ_E at a certain wavenumber to spread over the whole spectrum. It is this energy spread combined with a suitable vorticity forcing that appears to give rise to a nonlinear instability and the bifurcation leading to multiple steady-states. The experiments with the high resolution model have also confirmed that a suitably positioned vorticity forcing in a wave component enhances this bifurcation mechanism.

Another aspect of using one Fourier component to represent the orography which can be tested with a high resolution model, is to see whether multiple equilibria can be obtained with an isolated orographic ridge. Davey (1981) did an experiment of this type with an annular geometry and within a certain, rather narrow, range of the forcing parameter space, he obtained multiple steady-states. An example of two stable states can be found in Fig. 6. These states have the same characteristics as the blocked and zonal states described previously. It can also be seen from Fig. 6 that in the high amplitude wave state the waves generated on the leeward side of the orographic ridge are almost totally dissipated when the flow reaches the upwind side of the orographic ridge. The high amplitude wave state is thus not associated with a global resonance, the phenomenon is rather local in character. The resonance occurring is instead of the type where the Rossby wave generated by the orography has a phase speed which is such that it is stationary in the zonal flow which results.

5. OBSERVATIONAL STUDIES SUPPORTING A BIFURCATION MECHANISM FOR BLOCKING

The main conclusion that can be drawn from the bifurcation mechanism found in barotropic models is that the orography is necessary as a triggering mechanism in establishing the multiple steady-states. The implied application of the theory to atmospheric blocking can thus be tested by studying the effect of the orography on the atmospheric flow in connection with blocked flow situations. One parameter which reflects the influence of the orography on the barotropic component of atmospheric flow is the mountain torque. To furthermore couple observational evidence with the combination of orographic and wave vorticity forcing, an evaluation of the long wave forcing is needed. This forcing should be seen as the cumulative effect of the transient eddies on the mean flow rather than a direct thermal forcing.

To study the orographic factors influencing blocking action in the Atlantic and Pacific regions separately it is necessary to separate the torque contributions from the American and the Eur-Asian continents. It is primarily downstream from a mountain range that the orography may influence the flow pattern. The mountain torque parameter essentially reflects the surface pressure distribution around a mountain range and therefore it would be possible to separate the contributions from each continent by computing the torque for the eastern and western parts of the Northern Hemisphere separately. The separation line between eastern and western parts would then have to lie entirely within the oceanic regions. Computations of the mountain torque around complete latitute circles has previously been done by Oort and Bowman (1974). They presented monthly averaged results for a five year period including the anomalous winter of 1963. In January 1963 there was a well developed blocking ridge over the Atlantic region (see O'Connor, 1963) and for this month there was an exceptionally high mountain drag in midlatitudes. Recent calculations by Metz (private communication) have also shown a correlation between high values of the mountain drag (again around complete latitude circles) and high amplitudes of the geopotential over the eastern Atlantic area.

Separating the torque contributions from the North American and the Eur-Asian continents, Källén (1982) has shown that during one winter period (January-March, 1979) there is a correspondence between a mountain drag and the occurrence of a blocking ridge downstream of the mountain barrier.

The mountain drag, T_M, is defined as

$$T_M = - \int_A P_s \frac{\partial h}{\partial \lambda} a \cos\phi \, dS \qquad (9)$$

where p_s is the surface pressure, h the surface elevation above sea level, a the radius of the earth and ϕ, λ the latitudinal and longitudinal coordinates, respectively. The integration is carried out over two areas, one containing the North American continent and another containing the Eur-Asian continent. Both areas extend from the North Pole down to 30°N. The separation lines between the two areas lie entirely within oceanic regions and the torque contributions from both continents can thus be evaluated separately (Fig. 7). A time plot of the torque for the two areas is shown in Fig. 8. The time evolution of the mountain torque has been smoothed with a five day running time mean to remove the influence of short lived, travelling baroclinic eddies. These tend to give large variations in the torque on a time scale of one day.

The most prominent feature of the curves in Fig. 8 is the large variation of the torque on a time scale of about two weeks. Both curves tend to show sudden jumps between what appears to be fairly constant values, the jumps appearing over a time interval which is shorter than the time scale over which the torque is approximately constant. To investigate whether there is

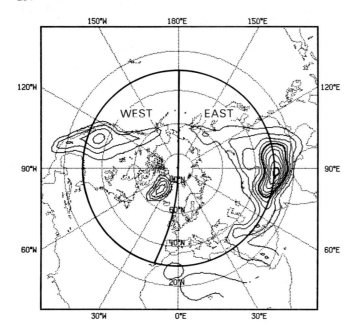

Fig. 7. Geographical map of the orographic field used to calculate the mountain torque. Contours are drawn with a 500 m interval. Thick lines indicate the boundaries of the western and eastern regions used for the calculation of the separate contributions to the mountain torque

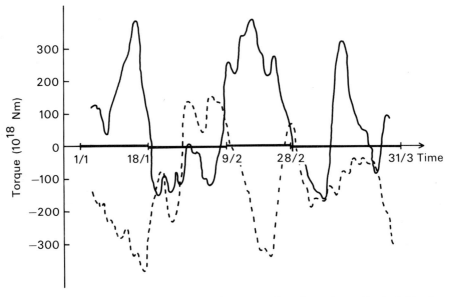

Fig. 8. Mountain torque for western (full line) and eastern (dashed line) parts of Northern Hemisphere ($30°N-90°N$) during January-March, 1979. The curves are calculated from 12 hourly FGGE data and a running 5-day mean time filter has been applied to smooth the curves. The time periods which are used as averaging periods for Fig. 9 are indicated on the horizontal time axis

any relation between the periods of low values of the torque (negative values of the torque imply a mountain drag) and the blocking events during January-March 1979, time mean maps of the 500 mb flow have been prepared for the time periods indicated in Fig. 8. The time periods are chosen to coincide with the periods between the jumps in Fig. 9. Also shown in Fig. 9 are areas with a high value of the time variability of the 500 mb surface height. The variability is calculated as the standard deviation of the 12-hourly values from the period average. A high variability indicates an intense eddy activity which can be coupled to strong baroclinic developments at mid-latitudes. A time plot of the variability averaged over regions upstream of the characteristic blocking regions is shown in Fig. 10. The variability in this figure is calculated as the standard deviation from a running seven day average of the 500 mb surface height. A high value of this quantity may thus be interpreted as an intense activity of eddies with a characteristic life time which is between 12 hours and seven days.

The time series has been divided into the following four periods.
- I.   1 Jan 1979 - 18 Jan 1979
- II.   18 Jan 1979 - 9 Feb 1979
- III. 9 Feb 1979 - 28 Feb 1979
- IV.   28 Feb 1979 - 31 Mar 1979

During periods I and III there are well developed ridges over the Pacific Ocean and the ridge in period III has many of the characteristics of a blocked flow. In the Atlantic region there is a well developed block during period II while there is also a block over Europe during period III. During period IV there is a predominantly zonal flow over the Atlantic-European region while over the Pacific there is a strong ridge in the poleward part of the region while the flow is zonal across the central Pacific Ocean. Going back to the plot of the mountain torque (Fig. 8), one may see that the Atlantic block during period II and the Pacific block during period III are

TIME AVERAGED 500 mb SURFACE

1-18 JANUARY 1979

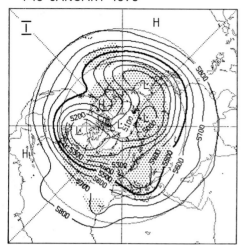

18 JANUARY - 9 FEBRUARY 1979

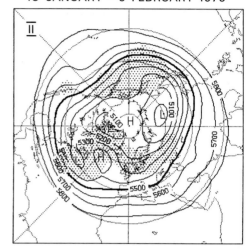

9-28 FEBRUARY 1979

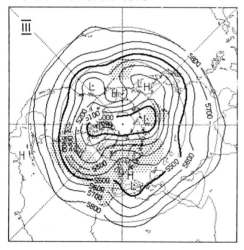

28 FEBRUARY - 31 MARCH 1979

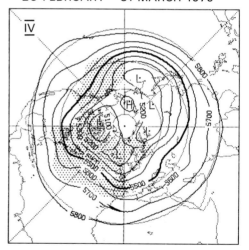

Fig. 9. Time mean maps of the 500 mb surface for the time periods indicated in Fig. 8. Areas where the variance is above 120^2 m^2 are hatched

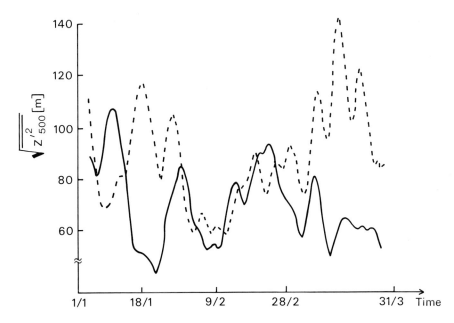

Fig. 10. Time evolution of variances as calculated from deviations of the 500 mb surface from a seven day running mean and spatially averaged over certain regions. All regions extend from 30°N to the North Pole and have the following longitudinal boundaries:

Full line, 135°E - 180°E (Kuroshio region of Pacific Ocean)

Dashed line, 45°W - 90°W (Gulfstream region of Atlantic Ocean)

coupled with low, negative values of the torque, i.e. a mountain drag. The Pacific ridge during period I is also associated with a high mountain drag over the Eur-Asian continent while the zonal flows over the Atlantic region during periods I and III and over the Pacific during period II are coupled with high, positive values of the torque. During the last period (IV) the torque from the North American continent shows considerable fluctuations and the flow over the Atlantic and European regions is predominantly zonal. The torque from the Eur-Asian continent is distinctly negative and there is a ridge extending northwards towards the polar regions over the Pacific Ocean. The occurrence of a high mountain drag thus has some correspondance with the appearance of a blocking high downstream of a continent.

The influence of the transient motion on the time mean flow is the mechanism which in the barotropic model is represented by a direct vorticity forcing. Evaluating this vorticity forcing from atmospheric data is difficult as discussed by Savijärvi (1978). Some attempts have been made at calculating the vorticity forcing for the time periods indicated in Fig. 8, but the results have generally been noisy and it has been difficult to see any clear pattern. Instead, the eddy activity has been evaluated in terms of the standard deviation of the 500 mb surface during the different time periods. From Fig. 9 it may be seen that the eddy activity upstream of a blocking region has some connection with the time periods defined earlier. Through Fig. 10 it may be seen that the Atlantic blocking during period II and to some extent the Pacific blocking during period III are coupled with a strong eddy activity in the beginning of those periods. However, as the blocking period continues there seems to be a decline in the activity of the eddies, especially during period II and in the Atlantic region. The decay of the block may therefore be coupled with the declining eddy activity, when the block has disappeared there is a renewed intensification of the eddy activity. This process is clearly baroclinic and the barotropic model of the previous sections cannot simulate it. It would be of interest to see if a simple baroclinic model could show these effects in the presence of orography. The reasoning does not hold with the period III block in the Pacific region, there the blocking is preceded by a very low eddy activity. On the geographical maps (Fig. 9) it may however be seen that there is a small region with a fairly high eddy activity just upstream of the block and it may be that the procedure of averaging the variance over a large area smooths this feature out. In any case, the geographical maps clearly show that the upstream flanks of the blocked regions do have intense eddy activity on the average and this is to be expected from the well known fact that in the Gulfstream and Kuroshio regions of the Atlantic and Pacific Oceans respectively, there is normally an intense baroclinic activity.

The ridge over Europe during period III, which on the daily 500 mb maps can be associated with blocking like patterns, is not connected with a high mountain drag over the American continent. This may be explained by the fact that the ridge is too far downstream from the orography to be significantly influenced by it and the blocking ridges may therefore develop due to some other mechanism. From the plot of the variance of the 500 mb surface (Fig. 7) it may however be noted that within the European area there is quite a large variability during this time period. This can be due to blocking like ridges moving across the area and not remaining stationary which also gives a smoothed ridge on a time averaged map. It may thus be a situation in which transient ridges develop, but because of the orientation of the large scale flow and the effect of the orography, the ridges cannot remain stationary to form a persistent block.

6. DISCUSSION

Investigations of the nonlinear properties of simplified atmospheric models have shown that a combination of orographic and vorticity forcing in barotropic, quasi-geostrophic models gives rise to a long wave instability and the development of multiple, stable equilibrium states. One of these stable states can be associated with a large amplitude wave response, while another has a dominating zonal flow and a less pronounced wave component. The large amplitude wave response (or blocked response) is close to a resonant flow configuration, where the wave response is almost in phase with the wave forcing. The zonal steady state is further removed from resonance and the response in the wave component is much weaker. Instead, due to the changed phase relationship between the wave and the orography, the zonal flow is more intense and the mountain drag is lower. These two types of equilibria can exist for the same values of all the forcing parameters, which one the flow choses is crucially dependent on the initial state of the flow in relation to the unstable steady-state.

The forcing parameters required in the barotropic model for the development of multiple equilibria, can be associated with the conditions present in the Northern Hemisphere during wintertime. A strong zonal flow and an intense baroclinic eddy activity off the eastern coasts of the two major continents can thus be linked with two possible types of response of the long waves in the atmosphere. Once the atmosphere has settled into one of these response types it is likely to remain there for an extended period of time due to the stability of the flow configuration. In one of the response types there is a well developed ridge downstream of a continent and this ridge can be associated with a blocked flow. To remain in this near resonant flow configuration it is also necessary that the eddy activity upstream of the blocking ridge is maintained to give an input of kinetic energy on the long waves. From the diagnostic studies of Källén (1982) it appears that this eddy activity steadily decreases during a blocking event and this may be the cause for the vanishing of the blocking ridge. A decreased eddy activity would, according to the barotropic mechanism put forward in Section 3, imply that the blocked steady-state vanishes (for a constant zonal forcing) and the flow is forced to settle into a zonal flow configuration. In Fig. 2 this may be visualized as the disappearance of curve II when the flow has settled into a state on the high index branch of curve II. At a critical value of the eddy forcing of the long waves the flow would be forced to transfer to a more zonal type of circulation.

A similar way in which cyclonic eddies and the orography may interact has been studied by Kalnay-Rivas and Merkine (1981). They investigated the effect of isolated vortices when periodically released upstream of an orographic ridge in a barotropic flow. A resonant, blocked type of response was found downstream of the orography when the time-averaged wave generated by the cyclonic eddy forcing was in phase with the orographically generated Rossby wave. This is the same type of reinforcing interaction which has been described in this paper, although Kalnay-Rivas and Merkine (1981) did not find any multiple equilibrium due to the nonlinear interactions. Instead they emphasized the effect of the nonlinearity in producing an Ω-shaped blocking pattern.

Recent investigations by Lau (1981) and Volmer et al (1981) on the behaviour of the GFDL and ECMWF general circulation models in long time integrations have interesting connections with the bifurcation mechanism discussed here. The data from the long integrations were analysed through an expansion into empirical orthogonal functions. Lau (1981) showed that he could find two characteristic types of wintertime circulations in the Northern Hemisphere, one with a predominantly zonal flow and another with a more pronounced meridional flow. Examining the variance during these two characteristic types of months, Lau (1981) was also able to show that the cyclone tracks over the Pacific and North Atlantic Oceans were very different during these months. During the months with a high zonal index the regions with a high variance extended across the oceans, while during the months with a low zonal index the variance was high only over the eastern parts of the oceans. It thus seems that the model has two different modes of circulation during wintertime in the Northern Hemisphere and the characteristic properties of these modes agree quite well with those of the two stable states found in a simple, low order, barotropic model.

References

Charney, J.G. and DeVore, J.G. 1979 Multiple flow equilibria in the atmosphere and blocking. J.Atmos.Sci., 36, 1205-1216.

Charney, J.G. and Strauss, D.M. 1980 Form-drag instability, multiple equilibria and propagating planetary waves in baroclinic, orographically forced, planetary wave systems. J.Atmos.Sci., 37, 1157-1176.

Davey, M.K. 1980 A quasi-linear theory for rotating flow over topography. Part I: Steady -plane channel. J.Fluid Mech., 99, 267-292.

Davey, M.K. 1981 A quasi-linear theory for rotating flow over topography. Part II: Beta-plane annulus. J.Fluid Mech., 103, 297-320.

Herring, J.R. 1963 Investigation of problems in thermal convection. J.Atmos.Sci.,20, 325-338.

Hoskins, B.J. 1973 Stability of the Rossby-Haurwitz wave. Quart.J.R.Met.Soc., 99, 723-745.

Hoskins, B.J. and Simmons, A.J. 1975 A multi-layer spectral model and the semi implicit method. Quart.J.R.Met.Soc., 101, 637-655.

Kalnay-Rivas,E. and Merkine, L.O. 1981 A simple mechanism for blocking. J.Atmos.Sci., 38, 2077-2091.

Källén,E. 1981 The nonlinear effects of orographic and momentum forcing in a low-order,barotropic model. J.Atmos.Sci., 38, in press.

Källén,E. 1982 Bifurcation properties of quasi-geostrophic, barotropic models and the relation to atmospheric blocking. Tellus, in press.

Lau,N.C. 1981 A diagnostic study of recurrent meteorological anomalies appearing in a 15-year simulation with a GFDL general circulation model. Mon.Wea.Rev., 109 , 2287-2311.

Lorenz, E.N. 1960 Maximum simplification of the dynamic equations. Tellus, 12, 243-254.

Marsden, J.E. and McCracken, M. 1976 The Hopf bifurcation and its applications. Springer Verlag, New York, pp.408.

O'Connor, J.F. 1963 The weather and circulation of January 1963. Mon.Wea.Rev., 91, 209-218.

Oort, A.H. and Bowman, H.D. 1974 A study of the mountain torque and its interannual variations in the Northern Hemisphere. J.Atmos.Sci., 31, 1974-1982.

Reinhold, Brian B., Pierrehumbert, Raymond T., 1982 "Dynamics of weather regimes: quasi-stationary waves and blocking", Mon.Wea.Rev., 110, 1105-1145.

Roads, J.O. 1980 Stable near-resonant states forced by orography in a simple baroclinic model. J.Atmos.Sci., 37, 2381-2395.

Saltzmann, B. 1970 Large-scale atmospheric energetics in wave-number domain. Rev. of Geophysics and Space Physics, 8, 289-302.

Savijärvi, H. 1978 The interaction of the monthly mean flow and large scale transient eddies in two different circulation types, Part II. Geophysica, 14, 207-229.

Steinberg, H.L., Wiin-Nielsen, A. and C.-H. Young 1971 On nonlinear cascades in large-scale atmospheric flow. J.Geophys.Res., 76, 8629-8640.

Trevisan, A. and Buzzi, A. 1980 Stationary response of barotropic weakly nonlinear Rossby waves to quasi-resonant orographic forcing. J.Atmos.Sci., 37, 947-957.

Volmer, J.P., Deque, M. and Jarraud, M. 1983 Large scale fluctuations in a long range integration of the ECMWF spectral model. Tellus, 35A, 173-188.

Wiin-Nielsen, A.C. 1979 Steady states and stability properties of a low order barotropic system with forcing and dissipation. Tellus, 31, 375-386.

Dynamically Stable Nonlinear Structures

C. E. Leith

1. INTRODUCTION

In recent years there has been increasing interest in the possibility that there may be dynamically stable nonlinear structures, such as solitons or modons, embedded in the otherwise quasi-two dimensional turbulent flow characterizing the large-scale behavior of the atmosphere. This possibility has important implications for predictability and prediction. It has been suggested, for example, by McWilliams (1980) that dipole blocking structures in the atmosphere may be modons.

Solitons and modons are special localized solutions of the nonlinear dynamics equations. Two aspects of these equations -- nonlinear interaction and linear dispersion -- might be expected to destroy any local structure. Together, however, they can balance each other and preserve certain structures in a surprisingly stable way.

Long (1964) found soliton solutions in a β-channel and Benney (1966) carried through a detailed perturbation calculation to derive a Korteweg-deVries (KdV) equation appropriate for Rossby waves. Solitons are exact solutions of the KdV equation, but the KdV equation is only an approximation to, say, the barotropic vorticity equation and is only formally valid for weak dispersion and weak nonlinearity. The KdV equation describes dependence in only one space dimension taken as longitude in these applications. Latitudinal dependence is determined by the constraints of meridional boundary conditions. Redekopp (1977) has worked out the theory of Rossby solitons in considerable detail.

Linear dispersion can also be induced by effects of bottom topography in a shallow water model of the ocean. The corresponding solitons have been examined for stability and practical numerical simulation by Malanotte Rizzoli (1980). In her simulations, she finds that soliton-like solutions persist and appear to be robust for conditions that greatly exceed the formal requirements of the perturbation theory.

An alternate construction of a localized solution of the barotropic vorticity equation was provided by Stern (1975). His modon solution is a dipole confined within a circle that is stationary with respect to uniform zonal flow. Modons are exact solutions rather than perturbation approximations but suffer from discontinuities at the circle boundary. Larichev and Reznik (1976) generalized Stern's modon and weakened the boundary discontinuity by attaching an exterior solution that decayed sufficiently rapidly to preserve the local nature of the modon. Such modons move with respect to a uniform zonal flow. Flierl et al. (1980) have generalized modons still further to equivalent barotropic and baroclinic cases and have shown that, having once constructed a dipole modon, monopole riders of great variety may be added. McWilliams (1980) has matched the parameters of an equivalent barotropic modon roughly to the observed characteristic of a dipole atmosphere blocking event observed over the North Atlantic Ocean in January 1963.

I shall describe in detail the construction of an equivalent barotropic modon in Section 2 and of its riders in Section 3. The spectral consequences of a rider vorticity discontinuity are examined in Section 4, and finally in Section 5 I shall summarize the results of some numerical modon experiments.

2. MODON STRUCTURE

We shall describe a localized modon solution for the equivalent barotropic vorticity equation

$$(\nabla^2 \psi - \alpha^2 \psi)_t + \beta \psi_x + J(\psi, \nabla^2 \psi) = 0 \tag{2.1}$$

which determines the evolution of the stream function ψ for the flow of shallow water of mean depth h on a β-plane with Coriolis coefficient $f = f_0 + \beta y$. Here α is the deformation wavenumber with

$$\alpha^2 = \frac{f_0^2}{gh} . \tag{2.2}$$

and g is an equivalent gravitational acceleration such that $(gh)^{1/2}$ is the speed of gravity waves. The Jacobian here is defined in the usual way as

$$J(\psi, \phi) = \psi_x \phi_y - \psi_y \phi_x \tag{2.3}$$

There exists in this case a potential vorticity

$$Z = f + \nabla^2 \psi - \alpha^2 \psi \tag{2.4}$$

in terms of which Eq. (2.1) may be rewritten

$$Z_t + J(\psi, Z) = 0 \tag{2.5}$$

displaying Z as conserved following the flow.

The linear Rossby wave solutions of Eq. (2.1) are given by eigenfunctions of ∇^2 such that

$$\nabla^2 \psi = -\lambda^2 \psi \tag{2.6}$$

For these the Jacobian term vanishes and Eq. (2.1) reduces to the linear equation

$$-(\alpha^2+\lambda^2)\psi_t + \beta\psi_x = 0 \tag{2.7}$$

describing waves propagating in the x-direction with velocity

$$c = -\frac{\beta}{\alpha^2+\lambda^2} \tag{2.8}$$

Since $0 < \lambda^2 < \infty$, c is bounded with $-\beta/\alpha^2 < c < 0$. Rossby waves are oscillatory in space like $\sin \lambda x$ and are not therefore localized solutions.

A localized solution must drop off rapidly away from some central region. As an outer solution with this property we take another eigenfunction of ∇^2 but one such that

$$\nabla^2 \psi = \mu^2 \psi \tag{2.9}$$

In particular we choose

$$\psi = A K_1(\mu r) \sin\theta \tag{2.10}$$

where K_1 is the modified Bessel function of the second kind of order 1. Again in the outer region the Jacobian vanishes and Eq. (2.1) reduces to Eq. (2.7) but with λ^2 replaced by $-\mu^2$. The outer solution (2.10) propagates therefore in the x-direction with velocity

$$c = -\frac{\beta}{\alpha^2-\mu^2} \tag{2.11}$$

Since $0 < \mu^2 < \infty$, the range of possible c values for localized solutions is $-\infty < c < -\beta/\alpha^2$ and $0 < c < \infty$, disjoint from the possible Rossby wave velocities of Eq. (2.8). In Eq. (2.10) r and θ are polar coordinates in a moving frame with, say,

$$r^2 = (x-ct)^2 + y^2$$
$$\sin\theta = y/r \tag{2.12}$$

To avoid the singularity in ψ at $r = 0$ given by Eq. (2.10), we introduce a smooth inner solution which we let join the outer one at a circle of radius $r = a$. We take as the inner solution for $r \le a$

$$\psi = BJ_1(\lambda r) \sin\theta - C r \sin\theta \tag{2.13}$$

where J_1 is the Bessel function of order 1.
The first term is again an eigenfunction of ∇^2 satisfying Eq. (2.6) and would by itself propagate in the x-direction with a velocity given

by Eq. (2.8). The second term, however, introduces a constant advecting velocity C. In order that the inner and outer propagation velocities be the same we must impose a velocity constraint

$$C = \beta \left[\frac{1}{\lambda^2} - \frac{\alpha^2 + \lambda^2}{\lambda^2(\alpha^2 - \mu^2)} \right] \qquad (2.14)$$

that determines the coefficient C for any choice of inner and outer wavenumbers, λ and μ.

We match the inner and outer solutions at $r = a$ by imposing as many continuity conditions as possible. From the continuity of ψ and ψ_r at $r = a$ we have

$$A \, K_1(\mu a) = B \, J_1(\lambda a) - Ca , \qquad (2.15)$$

and

$$A \, \mu a \, K_1'(\mu a) = B \lambda a \, J_1'(\lambda a) - Ca \qquad (2.16)$$

Let $\lambda a = \bar{\lambda}$, $\mu a = \bar{\mu}$. By subtraction we may eliminate the term Ca and find

$$A [K_1(\bar{\mu}) - \bar{\mu} K_1'(\bar{\mu})] = B [J_1(\bar{\lambda}) - \bar{\lambda} J_1'(\bar{\lambda})] \qquad (2.17)$$

Recursion relations for Bessel functions permit Eq. (2.17) to be written in simpler form as

$$A [\bar{\mu} K_2(\bar{\mu})] = B [\bar{\lambda} J_2(\bar{\lambda})] \qquad (2.18)$$

whence

$$\begin{aligned} A &= S [\bar{\mu} K_2(\bar{\mu})]^{-1} \\ B &= S [\bar{\lambda} J_2(\bar{\lambda})]^{-1} \end{aligned} \qquad (2.19)$$

The coefficient S is determined by Eq. (2.15) to be

$$S = Ca \left[\frac{J_1(\bar{\lambda})}{\bar{\lambda} J_2(\bar{\lambda})} - \frac{K_1(\bar{\mu})}{\bar{\mu} K_2(\bar{\mu})} \right]^{-1} \qquad (2.20)$$

The conditions imposed so far suffice to determine the amplitude coefficients A, B, and C for any choice of radius a and wavenumbers λ and μ. By Eq. (2.11) a choice of μ is equivalent to a choice of the overall propagation velocity c. Then the choice of λ determines C by Eq. (2.14). If we next choose a radius a then $\bar{\lambda}$ and $\bar{\mu}$ are defined, S is determined by Eq. (2.20) and finally A and B by Eqs. (2.19).

The most important continuity conditions have been satisfied, but we still have the freedom to choose λ for a given value of μ in such a way that the vorticity $\zeta = \nabla^2 \psi$ is also continuous at $r = a$.

The continuity condition for ζ at $r = a$ is

$$A \bar{\mu}^2 K_1(\bar{\mu}) = -B \bar{\lambda}^2 J_1(\bar{\lambda}) \tag{2.21}$$

which may be combined with Eqs. (2.19) to give

$$\frac{\bar{\lambda} J_1(\bar{\lambda})}{J_2(\bar{\lambda})} = - \frac{\bar{\mu} K_1(\bar{\mu})}{K_2(\bar{\mu})} \tag{2.22}$$

For any value of $\bar{\mu}$ the expression on the right is well defined and negative. Thus λ must be in those intervals of the $\bar{\lambda}$ range where J_1 and J_2 have opposite sign. We shall consider only the gravest such interval $(j_1^{(1)}, j_2^{(1)})$ where $\bar{\lambda}$ is smallest and the inner solution has the smoothest structure. The solid curve in Fig. 1 shows the mapping $\bar{\mu} \to \bar{\lambda}$ into this interval given by Eq. (2.22).

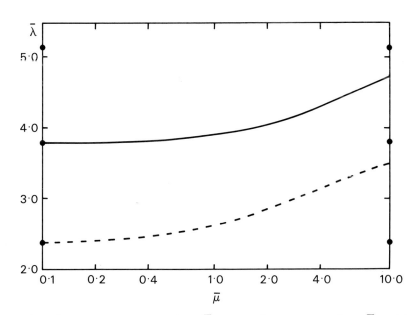

Fig. 1. Inner wavenumber $\bar{\lambda}$ vs. outer wavenumber $\bar{\mu}$ satisfying vorticity continuity conditions for a modon (solid) and a rider (dashed). Dots on the $\bar{\lambda}$ - axis delimit solution intervals

The modon so constructed is a localized vorticity dipole with an amplitude determined by its radius a and its velocity c. The required form of the second term in Eq. (2.13) imposes the dipole structure on the first term and on Eq. (2.10).

3. RIDER STRUCTURE

Once a modon has been constructed it is possible to add a rider to it and still have a localized propagating solution of Eq. (2.1). The simplest rider is a vorticity monopole. Let the rider stream function be indicated by ψ' to distinguish it from that of the modon; the total stream function will then be $\psi + \psi'$.

For the outer rider solution we choose a monopole solution of Eq. (2.9) with the same value of μ^2 thus for $r > a$

$$\psi' = D\, K_0(\mu r) \qquad (3.1)$$

The sum $\psi + \psi'$ is still an eigenvector of ∇^2 satisfying Eq. (2.9), and the velocity c is unchanged by the rider.

We take as the inner rider solution the monopole structure

$$\psi' = E\, J_0(\lambda r) + F \qquad (3.2)$$

with the same value of λ as for the modon. The first term is again an eigenfunction of ∇^2 satisfying Eq. (2.6). Since the constant F does not affect velocities, the velocity constraint of Eq. (2.14) is still appropriate and satisfied.

We match the inner and outer rider solutions at $r = a$ by imposing continuity conditions as for the modon. We find from continuity of ψ' and ψ'_r that

$$D = R\, [\bar{\mu} K_1(\bar{\mu})]^{-1}$$
$$E = R\, [\bar{\lambda} J_1(\bar{\lambda})]^{-1} \qquad (3.3)$$

and that

$$F = R \left[\frac{K_0(\bar{\mu})}{\bar{\mu} K_1(\bar{\mu})} - \frac{J_0(\bar{\lambda})}{\bar{\lambda} J_1(\bar{\lambda})} \right] \qquad (3.4)$$

where R is an arbitrary constant unconstrained by any imposed conditions. Once a modon has been constructed a monopole rider of an approximate shape may be added with arbitrary sign and amplitude.

Continuity of rider vorticity $\zeta' = \nabla^2 \psi'$ at $r = a$ leads through recursion relations to the condition

$$\frac{\bar{\lambda} J_2(\bar{\lambda})}{J_1(\bar{\lambda})} = \frac{\bar{\mu} K_2(\bar{\mu})}{K_1(\bar{\mu})} \qquad (3.5)$$

If this condition is satisfied then Eq. (2.20) determining the modon amplitude S becomes singular. The dashed curve in Fig. 1 shows the gravest solutions of Eq. (3.5). These must be avoided to preserve the modon structure.

A vorticity continuity condition may be imposed on the modon. It must not, however, be imposed instead on the rider, and the rider must have a vorticity discontinuity at r = a.

4. VORTICITY DISCONTINUITY

A discontinuity in a field tends to dominate its wavenumber spectrum at high wavenumbers and induce a characteristic power-law dependence. It is of interest to analyze this effect for the vorticity discontinuities that riders must have.

The essential aspect of the situation is revealed by considering the spectral transform of a vorticity function equal to a constant V for $r \leq a$ and vanishing for $r \leq a$. The two-dimensional fourier transform of a function f(r) depending only on r may be written as

$$\hat{f}(k_1,k_2) = \int_0^\infty r\, f(r) J_0(kr)\, dr = \hat{f}(k) \tag{4.1}$$

where $k_1 = k \sin\phi$, $k_2 = k \cos\phi$ are components of a two-dimensional wave vector. The transform $\hat{f}$ depends only on k as is to be expected from symmetry.

The two-dimensional power spectrum is proportional to the square of the amplitude

$$\phi(k_1,k_2) \propto \hat{f}^2(k) \tag{4.2}$$

The one-dimensional spectrum F(k) involves an integration over ϕ in wavevector space that introduces a factor k,

$$F(k) \propto k\, \hat{f}^2(k) . \tag{4.3}$$

In this case we have

$$f(r) = V \text{ for } r \leq a$$
$$= 0 \text{ for } r > a \tag{4.4}$$

so that

$$\hat{f}(k) = V \int_0^a r\, J_0(kr)\, dr$$

$$= V k^{-2} \int_0^{ka} x\, J_0(x)\, dx$$

$$= V a k^{-1} J_1(ka) \tag{4.5}$$

and

$$F(k) \propto V^2 a^2 k^{-1} J_1^2(ka) \tag{4.6}$$

Over the ensemble of radii and strengths this becomes

$$F(k) \propto k^{-1} <v^2 a^2 J_1^2(ka)> \qquad (4.7)$$

The evaluation of the bracketed average depends on knowing the joint probability distribution of v and a. In rough approximation we may assume that the average washes out the detailed oscillations of J_1^2 and leaves only its general dependence which is proportional to $(ka)^{-1}$ for ka >>1. Then the vorticity power spectrum becomes

$$F(k) \propto k^{-2} <v^2 a> \propto k^{-2}, \qquad (4.8)$$

and the associated kinetic energy spectrum is

$$E(k) = k^{-2} F(k) \propto k^{-4} \qquad (4.9)$$

Such a spectral contribution of rider discontinuities to the spectrum of two-dimensional turbulent motions would not dominate the inertial range spectrum that is proportional to k^{-3}.

5. NUMERICAL STUDIES

McWilliams et al. (1981) have carried out extensive gridpoint β-plane numerical studies of barotropic modons to determine the effects of limited resolution and of dissipative processes and the resistance of modons to various levels and scale of perturbations. They find modons to be remarkably robust and not easily destroyed by perturbations. In the resolution experiments, even with only five grid intervals per modon diameter, a modon-like structure persisted although with a characteristic velocity about one-half of the theoretical value.

Should modons or modon-like structures be relevant to weather and climate simulation, then the question of required model resolution becomes of considerable interest. Gridpoint methods are notoriously poor in inducing erroneous linear dispersion with associated errors in group velocity propagation of wave packets (Grotjahn and O'Brien, 1976). The slowing down of modons at low resolution observed by McWilliams et al. (1981) is likely to be a consequence, in part, of this kind of error. I have repeated their resolution experiments with a spectral transform β-plane model in which the linear terms are treated exactly with a linear recursion operator. Fig. 2 is taken from McWilliams et al. (1981) with circled points added to show my spectral transform results

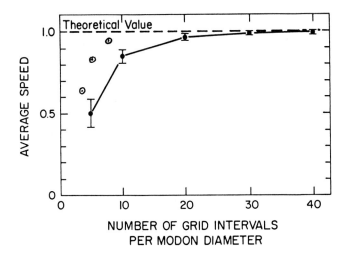

Fig. 2. Modon propagation speed as a function of resolution

With five spectral transform grid intervals per modon diameter, the velocity is still diminished but only by 15 percent. At coarse resolution, it appears that the modon-like structure tends to enlarge and in accordance with modon dynamics to slow down. These results suggest that even low-resolution spectral transform global models, say with rhomboidal 15 truncation, should be able to treat an atmospheric blocking structure of the sort studied by McWilliams (1980).

Many blocking events do not, however, have such a simple dipole structure and may be better fit with the help of riders. An important future task, therefore, is to carry out similar numerical experiments on the properties of modons with riders.

REFERENCES

Benney, D.J. 1966. Long nonlinear waves in fluid flows. J. Math. Phys. (Cambridge, Mass.), 45, 52-63.
Flierl, G.R. Larichev, V.D., McWilliams, J.C. and Reznik, G.M., 1980 The dynamics of baroclinic and barotropic solitary eddies. Dyn. Atmos. Oceans, 5, 1-41.
Grotjahn, R. and O'Brien, J.J. 1976. Some inaccuracies in finite differencing hyperbolic equations. Mon. Wea. Rev., 104, 180-194.
Larichev, V.D., and Reznik, G.M. 1976. A two-dimensional Rossby soliton-exact solution. In Rep. USSR Academy Sciences, 231 (5).

Long, R.R. 1964. Solitary waves in the westerlies. J. Atmos. Sci. 21, 197-200.

Malanotte Rizzoli, P. 1980. Solitary Rossby waves over variable relief and their stability. Dyn. Atmos. Oceans, 4, 261-294.

McWilliams, J.C. 1980. An application of equivalent modons to atmospheric blocking. Dyn. Atmos. Oceans, 5, 43-66.

McWilliams, J.C., Flierl, G.R., Larichev, V.D. and Reznik, G.M. 1981. Numerical studies of barotropic modons. Dyn. Atmos. Oceans, 5, 219-238.

Redekopp, L.G., 1977. On the theory of solitary Rossby waves. J. Fluid Mech., 82, 725-745.

Stern, M.E. 1975. Minimal properties of planetary eddies. J. Mar. Res. 33, 1-13.

Intense Atmospheric Vortices

Proceedings of the Joint Symposium (IUTAM/IUGG) held at Reading (United Kingdom) July 14-17, 1981

Editors: **L. Bengtsson, Sir J. Lighthill**

1982. 195 figures. XII, 326 pages
ISBN 3-540-11657-5

With contributions by F. K. Browand, M. Challa, D. R. Davies, R. P. Davies-Jones, Du Xing-yuan, M. P. Escudier, H. P. Evans, A. E. Gill, W. M. Gray, R. Hide, G. J. Holland, E. J. Hopfinger, J. B. Klemp, Y. Kurihara, L. M. Leslie, D. K. Lilly, B. A. Lugovtsov, T. Maxworthy, A. D. McEwan, K. V. Ooyama, R. P. Pearce, R. L. Pfeffer, R. Rotunno, R. C. Sheets, J. Simpson, R. K. Smith, J. T. Snow, R. E. Tuleya, Xiao Wen-jun, Xie An, Zhou Zi-Dong

The concept of vorticity is of central importance in fluid mechanics, and the change and variability of atmospheric flow is dominated by transient vortices of different time and space scales. Of particular importance are the most intense vortices, such as hurricanes, typhoons and tornadoes, which are associated with extreme and hazardous weather events of great concern to society.
This book examines the different mechanisms for vorticity intensification that operate in two different kinds of meteorological phenomena of great importance, namely the tropical cyclone and the tornado. The understanding of these phenomena has grown in recent years due to increased and improved surveillance by satellites and aircraft, as well as by numerical modelling and simulation, theoretical studies and laboratory experiments. The book summarizes these recent works with contributions from observation studies (from radio sonde data, aircraft, satellites and radars) and from studies concerning the physical mechanism of these vortices by means of theoretical, numerical or laboratory models. The book contains articles by the leading world experts on the meteorological processes and on the fundamental fluid dynamics mechanism, for vorticity intensification.

Springer-Verlag
Berlin
Heidelberg
New York
Tokyo

Eddies in Marine Science

Editor: **A. R. Robinson**

1983. 268 figures. XXV, 607 pages
(Topics in Atmospheric and Oceanographic Sciences)
ISBN 3-540-12253-2

Contents: Introduction. – Overview and Summary of Eddy Science. – Regional Kinematics, Dynamics, and Statistics. – Models. – Effects and Applications. – References. – Subject Index.

This book surveys the results of recent eddy research and explores its complications for ocean science and technology. It attempts a comprehensive review suitable for a wide audience of marine scientists. The investigation of eddy-current phenomena is rapidly advancing; however, many of the most fundamental dynamical questions of eddy dynamics are still not understood. The book therefore intends to conthribute to a global synthesis, to facilitate further research into eddy dynamics, and to encourage practical application. The knowledge of the physical science of eddies has important implications for biological, chemical, and geological oceanography, for modern ocean science and for practical activities in the sea including exploitation and management of the marine environment and its resources.

The book was produced under the auspices of the Working Group on Internal Dynamics of the Sea (WG-34) of the Scientific Comittee on Oceanic Research (SCOR) of the International Council of Scientific Unions.

Springer-Verlag
Berlin
Heideberg
NewYork
Tokyo